MATHEMATICS AS SIGN

WRITING SCIENCE

EDITORS Timothy Lenoir and Hans Ulrich Gumbrecht

MATHEMATICS AS SIGN

WRITING, IMAGINING, COUNTING

Brian Rotman

STANFORD UNIVERSITY PRESS
STANFORD, CALIFORNIA
2000

Stanford University Press
Stanford, California

Library of Congress Cataloging-in-Publication Data

Rotman, B. (Brian)
Mathematics as sign : writing, imagining, counting /
Brian Rotman.
p. cm.
Includes bibliographical references and index.
ISBN 0-8047-3683-9 (cloth : alk. paper) —
ISBN 0-8047-3684-7 (pbk. : alk. paper)
1. Mathematical notation. 2. Numeration. I. Title.
QA41.R67 2000
510—dc21 00-023291

This book is printed on acid-free, recycled paper.

Original printing 2000

Last figure below indicates year of this printing:
09 08 07 06 05 04 03 02 01 00

Printed in the United States of America

CONTENTS

PREFACE: WRITING, IMAGINING, COUNTING ix

1. Toward a Semiotics of Mathematics 1

2. Making Marks on Paper 44

3. How Ideal Are the Reals? 71

4. God Tricks; or, Numbers from the Bottom Up? 106

5. Counting on Non-Euclidean Fingers 125

NOTES 157

WORKS CITED 163

INDEX OF PERSONS 169

PREFACE: WRITING, IMAGINING, COUNTING

Mathematics is many things: the science of number and space; the study of pattern; an indispensable tool of technology and commerce; the methodological bedrock of the physical sciences; an endless source of recreational mind games; the ancient pursuit of absolute truth; a paradigm of logical reasoning; the most abstract of intellectual disciplines. In all of these and as a condition for their possibility, mathematics involves the creation of imaginary worlds that are intimately connected to, brought into being by, notated by, and controlled through the agency of specialized signs. One can say, therefore, that mathematics is essentially a symbolic practice resting on a vast and never-finished language—a perfectly correct but misleading description, since by common usage and etymology "language" is identified with speech, whereas one doesn't speak mathematics but writes it. Equally important, one doesn't write it as one writes or notates speech; rather, one "writes" in some other, more originating and constitutive sense.

Perhaps the most fundamental act of mathematical writing is the making of the Ur-marks or strokes in the form of patterns—1, 11, 111, 1111, 11111, and so on—that correspond to the actual and imagined activity of counting. The objectified results of these acts of pure semiotic repetition are the Ur-objects of mathematics we call integers. Asking, as one can hardly avoid doing when examining mathematical signs, what these ancient numerical objects are and what we might mean by them (*Was sind und Was sollen die Zahlen*, in Dedekind's famous formulation) entails negotiating the standard twentieth-century picture of numbers, and with it encountering the predominant philosophical belief of mathematicians, namely,

Platonism. That encounter is contentious and never very satisfactory. When probed, the philosophy quickly reveals its metaphysical-theological character, but it is so ingrained and does so much work (rhetorically, ideologically, and by way of a certain cognitive and disciplinary convenience) that disputing it—however polemically enjoyable and liberating—seems to get nowhere as far as its adherents are concerned. Nevertheless, as will appear, some disputation is at times necessary and inevitable.

The three themes that preoccupy us here, then, are mathematical writing, imagining, and counting: that is, the sense in which mathematics is a vast writing machine; how, in light of this sense, the activities of mathematical imagining and writing are interwoven; and what, in particular, are the ramifications of this interweaving for thinking—or, rather, rethinking—the status of the all-too-familiar things known as whole numbers.

MATHEMATICS AS SIGN

CHAPTER ONE

Toward a Semiotics of Mathematics

PREFACE

As the sign system whose grammar has determined the shape of Western culture's technoscientific discourse since its inception, mathematics is implicated, at a deeply linguistic level, in any form of distinctively intellectual activity. Indeed, the norms and guidelines of the "rational"—that is, the valid argument, definitional clarity, coherent thought, lucid explication, unambiguous expression, logical transparency, objective reasoning—are located in their most extreme, focused, and highly cultivated form in mathematics. The question this chapter addresses—what is the nature of mathematical language?—should therefore be of interest to semioticians and philosophers as well as mathematicians.

There are, however, certain difficulties inherent in attempting to address such disparate types of readers at the same time; it would be disingenuous not to acknowledge this at the outset.

Consider the mathematical reader. On the one hand, it is no accident that Peirce, whose writings created the possibility of the present chapter, was a mathematician; nor that I have practiced as a mathematician; nor that Hilbert, Brouwer, and Frege—the authors of the accounts of mathematics I shall dispute—were mathematicians. Mathematics is cognitively difficult, technical, abstract, and (for many) arid and defeatingly impersonal: one needs, it seems, to have been inside the dressing room in order to make much sense of the play. On the other hand, one cannot stay too long there if the play is not to disappear inside its own performance. In this respect, mathematicians confronted with the nature of their subject are no different from anybody else. The language that textual critics, for example, use to talk about criticism will be permeated by precisely those features—figures of ambiguity, polysemy, compression of meaning, subtlety and plurality of interpretation, rhetorical tropes, and so on—that these critics value in the texts they study. Likewise, mathematicians will create and respond to just those discussions of mathematics that ape what attracts them to their subject matter. Where textual critics literize their metalanguage, mathematicians mathematize theirs. And since for mathematicians the principal activity is proving new theorems, what they will ask of any description of their subject is: Can it be the source of new mathematical material? Does it suggest new notational systems, definitions, assertions, proofs? Now it is certainly the case that the accounts offered by Frege, Brouwer, and Hilbert all satisfied this requirement: each put forward a program that engendered new mathematics; each acted in and wrote the play and, in doing so, gave a necessarily truncated and misleading account of mathematics. Thus, if a semiotic approach to mathematics can be made to yield theorems and be acceptable to mathematicians, then it is unlikely to deliver the kind of exterior view of mathematics it promises. If it does not engender theorems, then mathematicians will be little interested in its project of redescribing their subject—the "queen of sciences"—via an explanatory formalism that (for them) is in a prescientific stage of arguing about its own fundamental terms. Since the account I have given is not slanted toward the creation of new mathematics, the chances of interesting mathematicians—let alone making a significant impact on them—look slim.

With readers versed in semiotics the principal obstacle is getting them sufficiently behind the mathematical spectacle to make sense of the project without losing them in the stage machinery. To this end I've kept the presence of technical discussion down to the absolute minimum. If I have been successful in this, then a certain dissatisfaction presents itself: the sheer semiotic skimpiness of the picture I offer. Rarely do I go beyond identifying an issue, clearing the ground, proposing a solution, and drawing a consequence or two. Thus, to take a single example, readers familiar with recent theories of narrative are unlikely to feel more than titillated by being asked to discover that the persuasive force of proofs, of formal arguments within the mathematical Code, are to be found in stories situated in the meta-Code. They would want to know what sort of stories, how they relate to each other, what their genres are, whether they are culturally and historically invariant, how what they tell depends on the telling, and so on. To have attempted to enter into these questions would have entailed the very technical mathematical discussions I was trying to avoid. Semioticians, then, might well feel they have been served too thin a gruel. To them I can only say that beginnings are difficult and that if what I offer has any substance, then others—they themselves, perhaps—will be prompted to cook it into a more satisfying sort of semiotic soup.

Finally, there are analytic philosophers. Here the difficulty is not that of unfamiliarity with the mathematical issues. On the contrary, no one is more familiar with them: the major thrust of twentieth-century analytic philosophy can be seen as a continuing response to the questions of reference, meaning, truth, naming, existence, and knowledge that emerged from work in mathematics, logic, and metamathematics at the end of the nineteenth century. And indeed, all the leading figures in the modern analytic tradition—from Frege, Russell, and Carnap to Quine, Wittgenstein, and Kripke—have directly addressed the question of mathematics in some way or other. The problem is rather that of incompatibility, a lack of engagement between forms of inquiry. My purpose has been to describe mathematics as a practice, as an ongoing cultural endeavor; and while it is unavoidable that any description I come up with will be riddled with unresolved philosophical issues, these are not—they *cannot* be if I am to get off the ground—my concern. Confronted, for example, with the debates and counterdebates contained in the elaborate secondary and tertiary literature on Frege, my response has been to avoid them and regard Frege's thought from a certain kind of semiotic scratch. So that if entering these debates is the only route to the attention of analytic philosophers, then the probability of engaging such readers seems not very great.

Obviously, I hope that these fears are exaggerated and misplaced and that there will be readers from each of these three academic specialisms prepared to break through what are only, after all, disciplinary barriers.

THREE SEMIOTIC ROUTES

My purpose here is to initiate the project of giving a semiotic analysis of mathematical signs; a project that, although implicit in the repeated references to mathematics in Peirce's writings on signs, seems so far not to have been carried out. Why mathematics, so obviously a candidate for semiotic attention, should have received so little of it will, I hope, emerge in these pages. Let me begin by presenting a certain obstacle, a difficulty of method, in the way of beginning the enterprise.

It is possible to distinguish, without being at all subtle about it, three axes or aspects of any discourse that might serve as an external starting point for a semiotic investigation of the code that underlies it. There is the referential aspect, which concerns itself with the code's secondarity, with the objects of discourse, the things that are supposedly talked about and referred to by the signs of the code; the formal aspect, whose focus is on the manner and form of the material means through which the discourse operates, its physical manifestation as a medium; and the psychological aspect, whose interest is primarily in those interior meanings that the signs of the code answer to or invoke. While all three of these axes can be drawn schematically through any given code, it is nonetheless the case that some codes seem to present themselves as more obviously biased toward just one of them. Thus, so-called representational codes, such as perspectival painting or realistically conceived literature and film, come clothed in a certain kind of secondarity; before all else they seem to be "about" some world external to themselves. Then there are those signifying systems, such as that of nonrepresentational painting, for example, where secondarity seems not to be in evidence, but where there is a highly palpable sensory dimension—a concrete visual order of signifiers—whose formal material status has a first claim on any semiotic account of these codes. And again, there are codes such as those of music and dance where what is of principal semiotic interest is how the dynamics of performance, of enacted gestures in space or time, are seen to be in the service of some prior psychological meanings assumed or addressed by the code.

With mathematics each of these external entry points into a semiotic account seems to be highly problematic: mathematics is an art that is practiced, not performed; its signs seem to be constructed—as we shall see—so as to sever their signifieds, what they are supposed to mean, from the real time and space within which their material signifiers occur; and the question of secondarity, of whether mathematics is "about" anything, whether its signs have referents, whether they are signs *of* something outside themselves, is precisely what one would expect a semiotics of mathematics to be in the business of discussing. In short, mathematics can offer only one of these familiar semiotic handles on itself—the referential route through an external world, the formal route through material signifiers, the psychological route through prior meanings—at the risk of begging the very semiotic issues requiring investigation.

To clarify this last point and put these three routes in a wider perspective, let me anticipate a discussion that can be given fully later in this chapter, only *after* a semiotic model of mathematics has been sketched. For a long time mathematicians, logicians, and philosophers who have written on the foundations of mathematics have agreed that (to put things at their most basic) there are really only three serious responses—mutually antagonistic and incompatible—to the question "What is mathematics?" The responses—formalism, intuitionism, and Platonism—run briefly as follows.

For the formalist, mathematics is a species of game, a determinate play of written marks that are transformed according to explicit unambiguous formal rules. Such marks are held to be without intention, mere physical inscriptions from which any attempt to signify, to mean, is absent: they operate like the pieces and moves in chess, and although they can be made to carry significance (representing strategies, for example), they function independently of such—no doubt useful but inherently posterior, after the event—accretions of meaning. Formalism, in other words, reduces mathematical signs to material signifiers that are, in principle, without signifieds. In Hilbert's classic statement of the formalist credo, mathematics consists of manipulating "meaningless marks on paper."

Intuitionists, in many ways the natural dialectical antagonists of formalists, deny that signifiers—whether written, spoken, or in any other medium—play any constitutive role in mathematical activity. For intuitionism mathematics is a species of purely mental construction, a form of internal cerebral labor, performed privately and in solitude within the individual—but cognitively universal—mind of the mathematician. If formalism characterizes mathematics as the manipulation of physical signifiers in the visible, intersubjective space of writing, then intuitionism (in Brouwer's formulation) sees it as the creation of immaterial signifieds within the Kantian—inner, a priori, intuited—category of time. And as the formalist reduces the signified to an inessential adjunct of the signifier, so the intuitionist privileges the signified and dismisses the signifier as a useful but theoretically unnecessary epiphenomenon. For Brouwer it was axiomatic that "mathematics is a languageless activity."

Last, and most important, since it is the orthodox position representing the view of all but a small minority of mathematicians, there is Platonism. For Platonists mathematics is neither a formal and meaningless game nor some kind of languageless mental construction, but instead a science, a public discipline concerned to discover and validate objective or logical truths. According to this conception, mathematical assertions are true or false propositions, statements of *fact* about some definite state of affairs, some objective reality, which exists independently of and prior to the mathematical act of investigating it. For Frege, whose logicist program is the principal source of twentieth-century Platonism, mathematics seen in this way was nothing other than an extension of pure logic. For his successors there is a separation: mathematical assertions are facts—specifically, they describe the properties of abstract collections (*sets*)—while logic is merely a truth-preserving form of inference which provides the means of proving that these descriptions are "true." Clearly, then, to the Platonist mathematics is a realist science, its symbols are symbols of certain real—prescientific—things, its assertions are consequently assertions about some determinate, objective subject matter, and its epistemology is framed in terms of what can be proved true *concerning* this reality.

The relevance of these accounts of mathematics to a semiotic project is twofold. First, to have persisted so long each must encapsulate, however partially, an important facet of what is felt to be intrinsic to mathematical activity. Certainly, in some undeniable but obscure way, mathematics seems at the same time to be a meaningless game, a subjective construction, and a source of objective truth. The difficulty is to extract these part-truths: the three accounts seem locked in an impasse which cannot be escaped from within the common terms that have allowed them to impinge on each other. As with the scholastic impasse created by nominalism, conceptualism, and realism—a parallel made long ago by Quine—the impasse has to be transcended. A semiotics of mathematics cannot, then, be expected to offer a synoptic reconciliation of these views; rather, it must attempt to explain—from a semiotic perspective alien to all of them—how each is inadequate, illusory, and undeniably attractive. Second, to return to our earlier difficulty of where to begin, each of these pictures of mathematics, although it is not posed as such, takes a particular theory-laden view about what mathematical signs are and are not; so that, to avoid a self-fulfilling circularity, no one of them can legitimately serve as a starting point for a semiotic investigation of mathematics. Thus, what we called the route through material signifiers is precisely the formalist obsession with marks, the psychological route through prior meaning comprehends intuitionism, and the route through an external world of referents is what all forms of mathematical Platonism require.

A SEMIOTIC MODEL OF MATHEMATICS

Where then can one start? Mathematics is an activity, a practice. If one observes its participants, then it would be perverse not to infer that for large stretches of time they are in engaged in a process of *communicating* with themselves and one another; an inference prompted by the constant presence of standardly presented formal written texts (notes, textbooks, blackboard lectures, articles, digests, reviews, and the like) being read, written, and exchanged, and of informal signifying activities that occur when they talk, gesticulate, expound, make guesses, disagree, draw pictures, and so on. (The relation between the formal and informal modes of communication is an important and interesting one to which I will return later; for the present, however, I want to focus on the written mathematical text.)

Taking the participants' word for it that such texts are indeed items in a communicative network, our first response would be to try to "read" them, to try to decode what they are about and what sorts of things they are saying. Pursuing this, what we observe at once is that any mathematical text is written in a mixture of words, phrases, and locutions drawn from some recognizable natural language together with mathematical marks, signs, symbols, diagrams, and figures that (we suppose) are being used in some systematic and previously agreed upon way. We will also notice that this mixture of natural and artificial signs is conventionally punctuated and divided up into what appear to be complete grammatical sentences; that is, syntactically self-contained units in which noun phrases ("all points on *x*," "the number *y*," "the first and second derivatives of *z*," "the theorem *alpha*," and so on) are systematically related to verbs ("count," "consider," "can be evaluated," "prove," and so on) in what one takes to be the accepted sense of connecting an activity to an object.

Given the problem of "objects," and of all the issues of ontology, reference, "truth," and secondarity that surface as soon as one tries to identify what mathematical particulars and entities such as numbers, points, lines, functions, relations, spaces, orderings, groups, sets, limits, morphisms, functors, and operators "are," it would be sensible to defer discussion of the interpretation of nouns and ask questions about the "activity" that makes up the remaining part of the sentence; that is, still trusting to grammar, to ask about verbs.

Linguistics makes a separation between verbs functioning in different grammatical moods; that is, between modes of sign use that arise, in the case of speech, from different roles which a speaker can select for himself and his hearer. The primary such distinction is between the indicative and the imperative.

The indicative mood has to do with asking for (interrogative case) or conveying (declarative case) information: "The speaker of a clause which has selected the indicative plus declarative has selected for himself the role of informant and for his hearer the role of informed" (Berry 1975, 166). For mathematics, the indicative governs all those questions, assumptions, and statements of information—assertions, propositions, posits, theorems, hypotheses, axioms, conjectures, and problems—that either ask for, grant, or deliver some piece of mathematical content, some putative mathematical fact such as "there are infinitely many prime numbers," "all groups with 7 elements are abelian," "$5 + 11 + 3 = 11 + 3 + 5$," "there is a continuous curve with no tangent at any point," "every even number can be written as the sum of two prime numbers," or, less obviously, those that might be said to convey metalingual information such as "assertion A is provable," "x is a counterexample to proposition P," "definition D is legitimate," "notational system N is inconsistent," and so on. In normal parlance, the indicative bundles information, truth, and validity indiscriminately together; it being equivalent to say that an assertion is "true," that it "holds," that it is "valid," that it is "the case," that it is informationally "correct," and so on. With mathematics it is necessary to be more discriminating: being "true" (whatever that is ultimately to mean) is not the same attribute of an assertion as being valid (that is, capable of being proved); conversely, what is informationally correct is not always, even in principle, susceptible of mathematical proof. The indicative mood, it seems, is inextricably tied up with the notion of mathematical proof. But proof in turn involves the idea of an argument, a narrative structure of sentences, and sentences can be in the imperative rather than the indicative.

According to the standard grammatical description, "the speaker of a clause which has chosen the imperative has selected for himself the role of controller and for his hearer the role of controlled. The speaker expects more than a purely verbal response. He expects some form of action" (Berry 1975, 166). Mathematics is so permeated by instructions for actions to be carried out, orders, commands, injunctions to be obeyed—"prove theorem T," "subtract x from y," "drop a perpendicular from point P onto line L," "count the elements of set S," "reverse the arrows in diagram D," "consider an arbitrary polygon with *k* sides," and similarly for the activities specified by the verbs "add," "multiply," "exhibit," "find," "enumerate," "show," "compute," "demonstrate," "define," "eliminate," "list," "draw", "complete," "connect," "assign," "evaluate," "integrate," "specify," "differentiate," "adjoin," "delete," "iterate," "order," "complete," "calculate," "construct," and so forth—that mathematical texts seem at times to be little more than sequences of instructions written in an entirely operational, exhortatory language.

Of course, mathematics is highly diverse, and the actions indicated even in this very incomplete list of verbs differ widely. Thus, depending on their context and their domain of application (algebra, calculus, arithmetic, topology, and so on), they display radical differences in scope, fruitfulness, complexity, and logical character: some (like "adjoin") might be finitary, while others (like "integrate") depend essentially on an infinite process; some (like "count") apply solely to collections, while others solely to functions or relations or diagrams, and still others (like "exhibit") apply to any mathematical entity; some can be repeated on the states or entities they produce, while others cannot. To pursue these differences would require technical mathematical knowledge that would be out of place in this project. It would also be beside the point: my focus on these verbs has to do not with the particular mathematical character of the actions they denote, but with differences between them—of an epistemological and semiotic kind—reflected in their grammatical status and, specifically, in their use in the imperative mood.

Corresponding to the linguist's distinction between inclusive imperatives ("Let's go") and exclusive imperatives ("Go"), there seems to be a radical split between types of mathematical exhortation: inclusive commands—marked by the verbs "consider," "define," "prove" and their synonyms—demand that speaker and hearer institute and inhabit a common world or that they share some specific argued conviction about an item in such a world; and exclusive commands—essentially the mathematical actions denoted by all other verbs—dictate that certain operations meaningful in an already shared world be executed.

Thus, for example, the imperative "consider a Hausdorff space" is an injunction to establish a shared domain of Hausdorff spaces; it commands its recipient to introduce a standard, mutually agreed upon ensemble of signs—symbolized notions, definitions, proofs, and particular cases that bring into play the ideas of topological neighborhood, limit point, a certain separability condition—in such a way as to determine what it means to dwell in the world of such spaces. By contrast, an imperative like "integrate the function f," for example, is mechanical and exclusive: it takes for granted that a shared frame (a world within the domain of calculus) has already been set, and asks that a specific operation relevant to this world be carried out on the function f. Likewise, the imperative "define . . ." (or equivalently, "let us define . . .") dictates that certain sign uses be agreed upon as the shared givens for some particular universe of discourse. Again, an imperative of the type "prove (or demonstrate or show) there are infinitely many prime numbers" requires its recipient to construct a certain kind of argument, a narrative whose persuasive force establishes a commonality between speaker and hearer with respect to the world of integers. By contrast, an imperative like "multiply the integer x by its successor" is concerned not to establish commonality of any sort, but to effect a specific operation on numbers.

One can gloss the distinction between inclusive and exclusive commands by observing that the familiar natural language process of forming nouns from verbs—the gerund "going" from the verb "go"—is not available for verbs used in the mood of the inclusive imperative. Thus, while exclusive commands can always (with varying degrees of artificiality, to be sure) be made to yield legitimate mathematical objects—"add" gives rise to an "adding" in the sense of the binary operation of addition, "count" yields a "counting" in the sense of a well-ordered binary relation of enumeration, and so on—such is not the case with inclusive commands: normal mathematical practice does not allow a "defining" or a "considering" or a "proving" to be legitimate objects of mathematical discourse. One cannot, in other words, prove results about or consider or define a "considering," a "proving," or a "defining" in the way that one can for an "adding," a "counting," and so on. (An apparent exception to this occurs in metamathematics, where certain sorts of definitions and proofs are themselves considered and defined and have theorems proved about them; but metamathematics is still mathematics—it provides no violation of what is being elaborated here.)

This grammatical line formulates the imperative in terms of speakers who dictate and hearers who carry out actions. But what is to be understood by "action" in relation to mathematical practice? What does the hearer—reader, recipient, addressee—actually do in responding to an imperative? Mathematics can be an activity whose practice is silent and sedentary. The only things mathematicians can be supposed to do with any certainty are scribble and think; they read and write inscriptions that seem to be inescapably attached to systematically meaningful mental events. If this is so, then whatever actions they perform must be explicable in terms of a scribbling/thinking amalgam. It is conceivable, as we have seen, to deny any necessary amalgamation of these two terms and to construe mathematics purely as scribbling (as entirely physical and "real": formalism's "meaningless marks on paper") or purely as thinking (as entirely mental and "imaginary": intuitionism's "languageless activity"); but to adopt either of these polemical extremes is to foreclose on any semiotic project whatsoever, since each excludes interpreting mathematics as a business of using those signifier/signified couples one calls *signs*. Ultimately, then, our object has to be to articulate what mode of signifying, of scribbling/thinking, mathematical activity is; to explain how, within this mode, mathematical imperatives are discharged; and to identify who or what semiotic agency issues and obeys these imperatives.

Leaving aside scribbling for the moment, let us focus on mathematical "thinking." Consider the imperative "consider a Hausdorff space." "Consider" means view attentively, survey, examine, reflect, and so on; the visual imagery here being part of a wider pattern of cognitive body metaphors, such as understand, comprehend, defend, grasp, or get the feel of an idea or thesis. Therefore, any attempt to explicate mathematical thought is unlikely to escape the net of such metaphors; indeed, to speak (as we did) of dwelling in a world of Hausdorff spaces is metaphorically to equate mathematical thinking with physical exploration. Clearly, such worlds are imagined, and the actions that take place within these worlds are imagined actions. Someone has to be imagining worlds and actions, and something else has to be performing these imaginary actions. In other words, someone—some subjective agency—is imagining itself to act. Seen in this way, mathematical thinking seems to have much in common with the making of self-reflective thought experiments. Such indeed was the conclusion Peirce arrived at:

> It is a familiar experience to every human being to wish for something quite beyond his present means, and to follow that wish by the question, "Should I wish for that thing just the same, if I had ample means to gratify it?" To answer that question, he searches his heart, and in so doing makes what I term an abstractive observation. He makes in his imagination a sort of skeleton diagram, or outline sketch of himself, considers what modifications the hypothetical state of things would require to be made in that picture, and then examines it, that is, observes what he has imagined, to see whether the same ardent desire is there to be discerned. By such a process, which is at bottom very much like mathematical reasoning, we can reach conclusions as to what *would* be true of signs in all cases. (Buchler 1940, 98)

Following the suggestion in Peirce's formulation, we are led to distinguish between sorts of mathematical agency: the one who imagines (what Peirce simply calls the "self" who conducts a reflective observation), which we shall call the *Subject* and the one who is imagined (the skeleton diagram and surrogate of this self), which we shall call the *Agent*. In terms of the distinction between imperatives, it is the Subject who carries out inclusive demands to "consider" and "define" certain worlds and to "prove" theorems in relation to these, and it is his Agent who executes the actions within such fabricated worlds, such as "count," "integrate," and so on, demanded by exclusive imperatives.

At first glance, the relation between Subject and Agent seems no more than a version of that which occurs in the reading of a road map, in which one propels one's surrogate, a fingertip model of oneself, around the world of roads imaged by the lines of the map. Unfortunately, the parallel is misleading, since the point of a road map is to represent real roads—real in the sense of being entities that exist prior to and independently of the map, so that an imagined journey by an agent is conceived to be (at least in principle) realizable. With mathematics the existence of such prior occurring "real" worlds is, from a semiotic point of view, problematic; if mathematical signs are to be likened to maps, then they are maps of purely imaginary territory.

In what semiotic sense is the Agent a skeleton diagram of the Subject? Our picture of the Subject is of a conscious—intentional, imagining—subject who creates a fictional self, the Agent, and fictional worlds within which this self acts. But such creation cannot, of course, be effected as pure thinking: signifieds are inseparable from signifiers: in order to create fictions, the Subject scribbles.

Thus, in response to the imperative "add the numbers in the list *S*," for example, he invokes a certain imagined world and—inseparable from this invocation—he writes down an organized sequence of marks ending with the mark to be interpreted as the sum of *S*. These marks are signifiers of signs by virtue of their interpretation within this world. Within this world "to add" might typically involve an infinite process, a procedure requiring that an infinity of actions be performed. This would be the case if, for example, *S* were the list of fractions 1, ½, ¼, ⅛, and so forth obtained by repeated halving. Clearly, in such a case, if "add" is to be interpreted as an action, it has to be an imagined action, one performed not by the Subject—who can only manipulate very small finite sequences of written signs—but by an actor imagined by the Subject. Such an actor is not himself required to imagine anything. Unlike the Subject, the Agent is not reflective and has no intentions: he is never called upon to "consider," "define," or "prove" anything, or indeed to attribute any significance or meaning to what he does; he is simply required to behave according to a prior pattern—do this then this then . . . —imagined for him by the Subject. The Agent, then, is a skeleton diagram of the Subject in two senses: he lacks the Subject's subjectivity in the face of signs; and he is free of the constraints of finitude and logical feasibility—he can perform infinite additions, make infinitely many choices, search through an infinite array, operate within nonexistent worlds—that accompany this subjectivity.

If the Agent is a truncated and idealized image of the Subject, then the latter is himself a reduced and abstracted version of the subject—let us call him the Person—who operates with the signs of natural language and can answer to the agency named by the "I" of ordinary nonmathematical discourse. An examination of the signs addressed to the Subject reveals that nowhere is there any mention of his being immersed in public historical or private durational time, or of occupying any geographical or bodily space, or of possessing any social or individualizing attributes. The Subject's psychology, in other words, is transcultural and disembodied. By writing its codes in a single tense of the constant present, within which addressees have no physical presence, mathematics dispenses entirely with the linguistic apparatus of deixis: unlike the Person, for whom demonstrative and personal pronouns are available, the Subject is never called upon to interpret any sign or message whose meaning is inseparable from the physical circumstances—temporal, spatial, cultural—of its utterance. If the Subject's subjectivity is "placed" in any sense, if he can be said to be physically self-situated, then his presence is located in and traced by the single point—the origin—required when any system of coordinates or process of counting is initiated; a replacement Herman Weyl once described as "the necessary residue of the extinction of the ego" (1949, 75).

I want now to bring this trio of semiotic actors—Agent, Subject, and Person—together and to display them as agencies that operate in relation to one another on different levels of the same mathematical process: namely, the centrally important process of mathematical proof. In the extract quoted above, Peirce likened what he called "reflective observation," in which a skeleton of the self takes part in a certain kind of thought experiment, to mathematical reasoning. For the Subject, reasoning is the process of giving and following proofs, of reading and writing certain highly specific and internally organized sequences of mathematical sentences—sequences intended to validate, test, prove, demonstrate, show that some particular assertion holds or is "true" or is "the case." Proofs are tied to assertions, and the semiotic status of assertions, as we observed earlier, is inextricable from the nature of proof. So, if we want to give a semiotic picture of mathematical reasoning as a kind of Peircean thought experiment, an answer has to be given to the question we dodged before: How, as a business having to do with signs, are we to interpret the mathematical indicative?

The answer I propose runs as follows. A mathematical assertion is a *prediction*, a foretelling of the result of performing certain actions upon signs. In making an assertion the Subject is claiming to know what would happen if the sign activities detailed in the assertion were to be carried out. Since the actions in question are ones that fall within the Subject's own domain of activity, the Subject is in effect laying claim to knowledge of his own future signifying states. In Peirce's phrase, the sort of knowledge being claimed is "what *would be* true of signs in all cases." Thus, for example, the assertion "2 + 3 = 3 + 2" predicts that if the Subject concatenates 1 1 with 1 1 1, the result will be identical to his concatenating 1 1 1 with 1 1. And more generally, "$x + y = y + x$" predicts that his concatenating any number of strokes with any other number will turn out to be independent of the order in which these actions are performed. Or, to take a different kind of example, the assertion that the square root of 2 is irrational is the prediction that whatever particular integers x and y are taken to be the result of calculating $x^2 - 2y^2$ will not be zero.

Obviously such claims to future knowledge need to be validated; the Subject has to persuade himself that *if* he performed the activities in question, the result would be as predicted. How is he to do this? In the physical sciences predictions are set against actualities: an experiment is carried out, and depending on the result, the prediction (or rather the theory giving rise to it) either is repudiated or receives some degree of confirmation. It would seem to be the case that in certain very simple cases such a direct procedure will work in mathematics. Assertions like "2 + 3 = 3 + 2" or "101 is a prime number" appear to be directly validatable: the Subject ascertains whether they correctly predict what he would experience by carrying out an experiment—surveying strokes or examining particular numbers—which delivers to him precisely that experience. But the situation is not, as we shall see below, that straightforward; and in any case, assertions of this palpable sort, although undoubtedly important in any discussion of the epistemological status of mathematical "truths," are not the norm. The Subject certainly cannot *by direct experiment* validate predictions like "$x + y = y + x$" or "the square root of 2 is irrational" unless he carries out infinitely many operations. Instead, as we have observed, he must act indirectly and set up an imagined experience—a thought experiment—in which not he but his Agent, the skeleton diagram of himself, is required to perform the appropriate infinity of actions.

By observing his Agent performing in his stead, by "reflective observation," the Subject becomes convinced—persuaded somehow by the thought experiment—that were *he* to perform these actions the result would be as predicted.

Now the Subject is involved in scribbling as well as thinking. The process of persuasion that a proof is supposed to achieve is an amalgam of fictive and logical aspects in dialectical relation to each other: each layer of the thought experiment (that is, each stage of the journey undertaken by the Agent) corresponds to some written activity, some manipulation of written signs performed by the Subject: so that, for example, in reading/writing an inclusive imperative the Subject modifies or brings into being a suitable facet of the Agent, and in reading/writing an exclusive imperative he requires this Agent to carry out the action in question, observes the result, and then uses the outcome as the basis for a further bout of manipulating written signs. These manipulations form the steps of the proof in its guise as a logical argument: any given step either is taken as a premise, an outright assumption about which it is agreed no persuasion is necessary, or is taken because it is a conclusion logically implied by a previous step. The picture offered so far, then, is that a proof is a logically correct series of implications that the Subject is persuaded to accept by virtue of the interpretation given to these implications in the fictive world of a thought experiment.

Such a characterization of proof is correct but inadequate. Proofs are arguments and, as Peirce forcefully pointed out, every argument has an underlying idea—what he called a "leading principle"—that converts what would otherwise be merely an unexceptionable sequence of logical moves into an instrument of conviction. The leading principle, Peirce argued, is distinct from the premise and the conclusion of an argument, and if added to these would have the effect of requiring a new leading principle and so on, producing an infinite regress in place of a finitely presentable argument. Thus, although it operates through the logical sequence that embodies it, it is neither identical nor reducible to this sequence; indeed, it is only by virtue of it that the sequence is an argument and not an inert, formally correct string of implications.

The leading principle corresponds to a familiar phenomenon within mathematics. Presented with a new proof or argument, the first question the mathematician (but not, see below, the Subject) is likely to raise concerns "motivation": he will in his attempt to understand the argument—that is, follow and be convinced by it—seek *the idea behind the proof.* He will ask for the story that is being told, the narrative through which the thought experiment or argument is organized. It is perfectly possible to follow a proof, in the more restricted, purely formal sense of giving assent to each logical step, without such an idea. If in addition an argument is based on accepted familiar patterns of inference, then its leading principle will have been internalized to the extent of being no longer retrievable: it is read automatically as part of the proof. Nonetheless a leading principle is always present—acknowledged or not—and attempts to read proofs in the absence of their underlying narratives are unlikely to result in the experience of felt necessity, persuasion, and conviction that proofs are intended to produce, and without which they fail to *be* proofs.

Now mathematicians—whether formalist, intuitionist, or Platonist—when moved to comment on this aspect of their discourse might recognize the importance of such narratives to the process of persuasion and understanding, but they are inclined to dismiss them, along with any other "motivational" or "purely psychological" or merely "aesthetic" considerations, as ultimately irrelevant and epiphenomenal to the real business of doing mathematics. What are we to make of this?

It is certainly true, as observed, that the leading principle cannot be part of the proof itself: it is not, in other words, addressed to the subject who reads proofs whom we have designated as the Subject. Indeed, the underlying narrative could not be so addressed, since it lies outside the linguistic resources mathematics makes available to the Subject. We might call the total of all these resources the mathematical *Code* and mean by this the discursive sum of all legitimately defined signs and rigorously formulated sign practices that are permitted to figure in mathematical texts. At the same time let us designate by the *meta-Code* the penumbra of informal, unrigorous locutions within natural language involved in talking about, referring to, and discussing the Code that mathematicians sanction. The Subject is the subject pertaining to the Code—the one who reads/writes its signs and interprets them by imagining experiments in which the actions inherent in them are performed by his Agent. We saw earlier that mathematicians prohibit the use of any deictic terms from their discourse; from which it follows that no description of himself is available to the Subject within the Code. Although he is able to imagine and observe the Agent as a skeleton diagram of himself, he cannot—within the vocabulary of the Code—articulate his relation to that Agent. He knows the Agent is a simulacrum of himself, but he cannot talk about his knowledge. Precisely in the articulation of this relation lies the semiotic source of a proof's persuasion: the Subject can be persuaded by a thought experiment designed to validate a prediction about his own actions only if he appreciates the resemblance—for the particular mathematical purpose at hand—between the Agent and himself. It is the business of the underlying narrative of a proof to articulate the nature of this resemblance. In short, the idea behind a proof is situated in the meta-Code; it is not the Subject himself who can be persuaded by the idea behind a proof, but the Subject in the presence of the Person, the natural language subject of the meta-Code for whom the Agent as a simulacrum of the Subject is an object of discourse.

What then, to return to the point above, is meant by the "real business" of doing mathematics? In relation to the discussion so far, one can say this: if it is insisted that mathematical activity be described solely in terms of manipulations of signs within the Code, thereby restricting mathematical subjectivity to the Subject and dismissing the meta-Code as an epiphenomenon, a domain of motivational and psychological affect, then one gives up any hope of a semiotic view of mathematical proof able to give a coherent account—in terms of sign use—of how proofs achieve conviction.

There are two further important reasons for refusing to assign to the meta-Code the status of mere epiphenomenon. The first concerns the completion of the discussion of the indicative. As observed earlier, there are, besides the assertions within the Code that we have characterized as predictions, other assertions of undeniable importance that must be justified in the course of normal mathematical practice. When mathematicians make assertions like "definition *D* is well-founded," or "notation system *N* is coherent," they are plainly making statements that require some sort of justification. Equally plainly, such assertions cannot be interpreted as predictions about the Subject's future mathematical experience susceptible to proof via a thought experiment. Indeed, indicatives such as "assertion *A* is provable" or "x is a counterexample to *A*," where *A* is a predictive assertion within the Code, cannot themselves be proved mathematically without engendering an infinite regress of proofs. It would seem that such metalingual indicatives—which of course belong to the meta-Code—admit "proof" in the same way that the proof of the pudding is in the eating: one justifies the statement "assertion *A* is provable" by exhibiting a proof of *A*. The second reason for treating the meta-Code as important to a semiotic account of mathematics relates to the manner in which mathematical codes and sign usages come into being, since it can be argued (although I will not do so here) that Code and meta-Code are mutually constitutive, and that a principal way in which new mathematics arises is through a process of catachresis—that is, through the sanctioning and appropriation of sign practices that occur in the first place as informal and unrigorous elements, in a merely descriptive, motivational, or intuitive guise, within the meta-Code.

The model I have sketched has required us to introduce three separate levels of mathematical activity corresponding to the sublingual imagined actions of an Agent, the lingual Coded manipulations of the Subject, and the metalingual activities of the Person, and then to describe how these agencies fit together. Normal mathematical discourse does not present itself in this way; it speaks only of a single unfractured agency, a "mathematician," who simply "does" mathematics. To justify the increase in complexity and artificiality of its characterization of mathematics, the model has to be useful; its picture of mathematics ought to illuminate and explain the attraction of the three principal ways of regarding mathematics we alluded to earlier.

FORMALISM, INTUITIONISM, PLATONISM

We extracted the idea of proof as a kind of thought experiment from Peirce's general remarks on reflective observation; we might have gotten a later and specifically mathematical version straight from Hilbert, from his formalist conception of metalogic as amounting to those "considerations in the form of *thought-experiments* on objects, which can be regarded as concretely given" (Hilbert and Bernays 1934, 20). But this would have sidetracked us into a description of Hilbert's metamathematical program, which it was designed to serve.

The object of this program was to show by means of mathematical reasoning that mathematical reasoning was consistent, that it was incapable of arriving at a contradiction. In order to reason about reasoning without incurring the obvious circularity inherent in such an enterprise, Hilbert made a separation between the reasoning that mathematicians use—which he characterized as formal manipulations, finite sequences of logically correct deductions performed on mathematical symbols—and the reasoning that the metamathematician would use, the metalogic, to show that this first kind of reasoning was consistent. The circularity would be avoided, he argued, if the metalogic was inherently safe and free from the sort of contradiction that threatened the object logic about which it reasoned. Since the potential source of contradiction in mathematics was held to be the occurrence of objects and processes *interpreted* by mathematicians to be infinitary, the principal requirement of his metalogic was that it be finitary and that it avoid interpretations—that is, that it attribute no meanings to the subject matter about which it reasoned. Hilbert's approach to mathematics, then, was to ignore what mathematicians thought they meant or intended to mean and, instead, to treat it as a formalism, as a system of meaningless written marks finitely manipulated by the mathematician according to explicitly stated formal rules. It was to this formalism that his metalogic, characterized as thought experiment on things, was intended to apply.

The first question to ask has to concern the "things" that are supposed to figure in thought experiments: What are these concretely given entities about which thought takes place? The formalist answer—objects concretely given as visible inscriptions, as definite but meaningless written marks—would have been open to the immediate objection that meaningless marks, while they can undoubtedly be manipulated and subjected to empirical (that is, visual) scrutiny, are difficult to equate to the sort of entities that figure in the finitary arithmetical assertions that form the basis for distinguishing experiments from *thought* experiments. Thus, in order to validate a finitary assertion like "2 + 3 = 3 + 2," the formalist mathematician supposes that a *direct* experiment is all that is needed: the experimenter is convinced that concatenating "11" and "111" is the same as concatenating "111" and "11" through the purely empirical observation that in both cases the result is the assemblage of marks "11111." But such an observation is a completely empirical validation—a pure *ad oculo* demonstration free of any considerations of meaning—only if the mathematical mark "1" is purely and simply an *empirical* mark, a mark all of whose significance lies in its visible appearance; and this, as philosophical critics of formalism from Frege onward have pointed out, is manifestly not the case. If it were, then arithmetical assertions would lack the generality universally ascribed to them; they would be about the physical perceptual characteristics of particular inscriptions—their exact shape, color, and size; their durability; the depth of their indentation on the page; the exact identity of one with another; and so on—and would need reformulating and revalidating every time one of these variables altered—a conclusion that no formalist of whatever persuasion would want to accept.

In fact, regardless of any considerations of intended meaning, the mathematical symbol "1" cannot be identified with a mark at all. In Peirce's terminology "1" is not a token (a concretely given visible object), but rather a *type*, an abstract pattern of writing, a general form of which any given material inscription is merely a perceptual instance. For us the mathematical symbol "1" is an item of thinking/scribbling, a *sign*, and we can now flesh out one aspect of our semiotic model by elaborating what is to be meant by this. Specifically, we propose: the sign "1" has for its signifier the type of the mark "1" used to notate it, and for its signified that relation in the Code, which we have yet to explicate, between thought and writing accorded to the symbol "1" by the Subject.

This characterization of "1" has the immediate consequence that mathematical signifiers are themselves dependent on some prior signifying activity, since types as entities with meaningful attributes—abstract, general, unchanging, permanent, exact, and so on—can come into being and operate only through the semiotic separation between real and ideal marks. Now neither the creation nor recognition of this ideality is the business of what we have called the Subject: it takes place before his mathematical encounter with signs. Put differently, the Subject assumes this separation but is not, and cannot be, called upon to mention it in the course of interpreting signs within the Code; it forms no part of the meaning of these signs insofar as this meaning is accessible to him as the addressee of the Code. On the contrary, it is the Person, operating from a point exterior to the Subject, who is responsible for this ideality; for it is only in the meta-Code, where mathematical symbols are discussed *as signs*, that any significance can be attributed to the difference between writing tokens and using types.

We can apply this view of signs to the question of mathematical "experience" as it occurs within formalism's notion of a thought experiment. The pressure for inserting the presence of an Agent into our model of mathematical activity came from the fact that whatever finitary actions upon signs the Subject might *in fact* be able to carry out, such as concatenating "11" and "111," these were the exception; for the most part the actions that he was called upon to perform (such as evaluating $x + y$ for arbitrary integers x and y) could be carried out only *in principle*, and for these infinitary actions he had to invoke the activity of an Agent. I suggested earlier that for the former sort of actions there appeared to be no need for him to invoke an Agent and a thought experiment; that the Subject might experience for himself by direct experiment the validity of assertions such as "2 + 3 = 3 + 2." This idea that (at least some) mathematical assertions are capable of being directly "experienced" is precisely what formalists—interpreting experience as meaning visual inspection of objects in space—claim. One needs therefore to be more specific about the meaning of direct mathematical "experience." From what has been said so far, no purely physical manipulation of purely physical marks can, of itself, constitute mathematical persuasion. The Subject manipulates types: he can be persuaded that direct experiments with tokens constitute validations of assertions like "2 + 3 = 3 + 2" only by appealing to the relationship between tokens and types, to the way tokens stand in place of types. And this he, as opposed to the Person, cannot do. The upshot, from a semiotic point of view, is that in no case can thought experiments be supplanted by direct experimentation; an Agent is always required. Validating finitary assertions is no different from validating assertions in general, where it is the Person who, by being able to articulate the relation between Subject and Agent within a thought experiment, is persuaded that a prediction about the Subject's future encounter with signs is to be accepted.

This way of seeing matters does not dissolve the difference between finitary and nonfinitary assertions. Rather, it insists that the distinction—undoubtedly interesting, but for us limitedly so in comparison to its centrality within the formalist program—between mathematical actions executable *in fact* and *in principle* makes sense only in terms of what constitutes mathematical persuasion, which in turn can be explicated only by examining the possible relationships between Agent, Subject, and Person that mathematicians are prepared to countenance as legitimate. (In reified form: How big can a "small" concretely surveyable collection of marks be before it becomes "large" and unsurveyable?)

Hilbert's program for proving the consistency of mathematics through a finitary metamathematics was, as is well known, brought to an effective halt by Gödel's theorem. But this refutation of what was always an overambitious project does not demolish formalism as a viewpoint; nor, without much technical discussion outside the scope of this chapter, can it be made to shed much light on how formalism's inadequacies arise. Thus, from a *semiotic* point of view the problems experienced by formalism can be seen to rest on a chain of misidentifications. By failing to distinguish between tokens and types, and thereby mistaking items possessing significance for pre-semiotic "things," formalists simultaneously misdescribe mathematical reasoning as syntactical manipulation of meaningless marks and metamathematical reasoning as thought experiments that theoreticalize these manipulations—in the sense that the formalists' "experiment," of which their thought experiment is an imagined theoretical version, is an entirely empirical process of checking the perceptual properties of visible marks. As a consequence, the formalist account of mathematical *agency*—which distinguishes merely between a "mathematician" who manipulates mathematical symbols as if they were marks and a "metamathematician" who reasons about the results of this manipulation—is doubly reductive of the picture offered by the present model, since at one point it shrinks the role of Subject to that of Agent and at another manages to absorb it into that of Person.

If formalism projects the mathematical amalgam of thinking/scribbling onto a plane of formal scribble robbed of meaning, intuitionism projects it onto a plane of thought devoid of any written trace. Each bases its truncation of the sign on the possibility of an irreducible mathematical "experience" that is supposed to convey by its very unmediated directness what it takes to be essential to mathematical practice: formalism, positivist and suspicious in a behaviorist way about mental events, has to locate this experience in the tangible written product, surveyable and "real"; intuitionism, entirely immersed in Kantian apriorism, identifies the experience as the process, the invisible unobservable construction in thought, whereby mathematics is created.

Brouwer's intuitionism, like Hilbert's formalism, arose as a response to the paradoxes of the infinite that emerged at the turn of the twentieth century within the mathematics of infinite sets. Unlike Hilbert, who had no quarrel with the Platonist conception of such sets and whose aim was to leave mathematics as it was by providing a post facto metamathematical justification of its consistency in which all existing infinitary thought would be legitimated, Brouwer attacked the Platonist notion of infinity itself and argued for a root-and-branch reconstruction from within in which large areas of classically secure mathematics—infinitary in character but having no explicit connection to any paradox—would have to be jettisoned as lacking any coherent basis in thought and, therefore, meaningless. The problem, as Brouwer saw it, was the failure on the part of orthodox—Platonist-inspired—mathematics to separate what for him are the proper objects of mathematics (namely, constructions in the mind) from the secondary aspect of these objects, the linguistic apparatus that mathematicians might use to describe and communicate about in words the features and results of any particular such construction. Confusing the two allowed mathematicians to believe, Brouwer argued, that linguistic manipulation was an unexceptional route to the production of new mathematical entities. Since the classical logic governing such manipulation has its origins in the finite states of affairs described by natural language, the confusion results in a false mathematics of the infinite: verbiage that fails to correspond to any identifiable mental activity, since it allows forms of inference that make sense only for finite situations, such as the law of excluded middle or the principle of double negation, to appear to validate what are in fact illegitimate assertions about infinite ones.

Thus, the intuitionist approach to mathematics, like Hilbert's scheme for metamathematics, insists that a special and primary characteristic of logic lies in its appropriateness to finitary mathematical situations. And although they accord different functions to this logic—for Hilbert it has to validate finitary metamathematics in order to secure infinitary mathematics, while for Brouwer it is an after-the-fact formalization of the principles of correct mental constructions finitary and infinitary—they each require it to be convincing: formalism grounding the persuasive force of its logic in the empirical certainty of *ad oculo* demonstrations, intuitionism being obliged to ground it, as we shall see, in the felt necessity and self-evidence of the mathematician's mental activity.

If mathematical assertions are construed Platonistically, as unambiguous, exact, and precise statements of fact, propositions true or false about some objective state of affairs, Brouwer's rejection of the law of excluded middle (the principle of logic that declares that an assertion either *is* the case or *is not* the case) and his rejection of double negation (the rule that not being not the case is the same as being the case) seem puzzling and counterintuitive: about the particulars of mathematics conceived in this exact and determinate way there would appear to be no middle ground between truth and falsity, and no way of distinguishing between an assertion and the negation of the negation of that assertion. Clearly, truth and falsity of assertions will not mean for intuitionists—if indeed they are to mean anything at all for them—what they do for orthodox mathematicians; and, since Platonistic logic is founded on truth, intuitionists cannot be referring to the orthodox process of deduction when they talk about the validation of assertions.

For the intuitionist, an assertion is a claim that a certain mental construction has been carried out. To validate such claims the intuitionist must either exhibit the construction in question or, less directly, show that it *can* be carried out by providing an effective procedure, a finite recipe, for executing it. This effectivity is not an external requirement imposed on assertions after they have been presented, but rather is built into the intuitionist account of the logical connectives through which assertions are formulated. From this internalized logic, principles such as the laws of excluded middle, double negation, and so on are excluded. Thus, in contrast to the Platonist validation of an existential assertion (x exists if the assumption that it doesn't leads to a contradiction), for the intuitionist x can be shown to exist only by exhibiting it, or by showing effectively how to exhibit it. Again, to validate the negation of an assertion A, it is not enough—as it is for the Platonist—to prove the existence of a contradiction issuing from the assumption A; the intuitionist must exhibit or show how one would exhibit the contradiction when presented with the construction claimed in A. Likewise for implication: to validate "A implies B," the intuitionist must provide an effective procedure for converting the construction claimed to have been carried out in A into the construction being claimed in B.

Clearly, the intuitionist picture of mathematical assertions and proofs depends on the coherence and acceptability of what it means by an effective procedure and (inseparable from this) on the status of claims that mental constructions have been or can be carried out. Proofs, validations, and arguments, in order to "show" or "demonstrate" a claim, have before all else to be convincing; they need to persuade their addressees to accept what is claimed. Where, then, in the face of Brouwer's characterization that "intuitionist mathematics is an essentially languageless activity of the mind having its origin in the perception of a move in time" (1952, 141), with its relegation of language—that is, all mathematical writing and speech—to an epiphenomenon of mathematical activity, a secondary and (because it is after the fact) theoretically unnecessary business of mere description, are we to locate the intuitionist version of persuasion? The problem is fundamental. Persuading, convincing, showing, and demonstrating are discursive activities whose business it is to achieve intersubjective agreement. But for Brouwer, the *inter*subjective collapses into the subjective: there is only a single cognizing subject privately carrying out constructions—sequences of temporally distinct moves—in the Kantian intuition of time. This means that, for the intuitionist, conviction and persuasion appear as the possibility of a replay, a purely mental reenactment within this one subjectivity: perform this construction in the inner intuition of time you share with me, and you will—you must—experience what I claim to experience.

Validating assertions by appealing in this way to felt necessity, to what is supposed to be self-evident to the experiencing subject, goes back to Descartes's cogito, to which philosophers have raised a basic and (this side of solipsism) unanswerable objection: what is self-evidently the case for one may not be self-evident—or worse, may be self-evidently not the case—for another; so that, since there can be no basis other than subjective force for choosing between conflicting self-evidence, what is put forward as a process of rational validation intended to convince and persuade is ultimately no more than a refined reiteration of the assertion it claims to be validating.

That intuitionism should be unable to give a coherent account of persuasion is what a semiotic approach which insists that mathematics is a business with and about *signs*, conceived as public, manifest amalgams of scribbling/thinking, would lead one to believe. The inability is the price intuitionism pays for believing it possible to first separate thought from writing and then demote writing to a description of this prior and languageless—presemiotic—thought. Of course, this is not to assert that intuitionism's fixation on thinking to the exclusion of writing is not useful or productive; within mathematical practice, by providing an alternative to classical reasoning, it has been both. And indeed, insofar as a picture of mathematics-as-pure-thought is possible at all, intuitionism in some form or other could be said to provide it.

From a semiotic viewpoint, however, any such picture cannot avoid being a metonymic reduction, a *pars pro toto* that mistakes a part—the purely mental activities that seem undeniably to accompany all mathematical assertions and proofs—for the whole writing/thinking business of manipulating signs and in so doing makes it impossible to recognize the distinctive role played by signifiers in the creation of mathematical meaning. Far from being the written traces of a language that merely *describes* prior mental constructions appearing as presemiotic events accessible only to private introspection, signifiers mark signs that are *interpreted* in terms of imagined actions which themselves have no being independent of their invocation in the presence of these very signifiers. In this dialectic relation between scribbling and thinking, whereby each creates what is necessary for the other to come into being, persuasion—as a tripartite activity involving Agent, Subject, and Person within a thought experiment—has to be located.

These remarks about formalism and intuitionism are intended to serve not as philosophical critiques of their claims about the nature of mathematics but as means of throwing into relief the contrasting claims of our semiotic model. From the point of view of mainstream mathematical practice, moreover, the formalist and intuitionist descriptions of mathematics are of minor importance. True, formalism's attempt to characterize finitary reasoning is central to metamathematical investigations such as proof theory, and intuitionistic logic is at the front of any constructivist examination of mathematics; but neither exerts more than a marginal influence on how the overwhelming majority of mathematicians regard their subject matter. When they pursue their business mathematicians do so neither as formalist manipulators nor as solitary mental constructors, but rather as scientific investigators engaged in publicly discovering objective truths. And they see these truths through Platonistic eyes: eternal verities, objective irrefutably-the-case descriptions of some timeless, spaceless, subjectless realm of abstract "objects."

Although the question of the nature of these Platonic objects—what are numbers?—can be made as old as Western philosophy, the version of Platonism that interests us (namely, the prevailing orthodoxy in mathematics) is a creation of nineteenth-century realism. And since our focus is semiotic and not philosophical, our primary interest is in the part played by a realist conception of *language* in forming and legitimizing present-day mathematical Platonism.

For the realist, language is an activity whose principal function is that of *naming*: its character derives from the fact that its terms, locutions, constructions, and narratives are oriented outward, that they point to, refer to, denote some reality outside and prior to themselves. They do this not as a by-product, consequentially on some more complex signifying activity, but essentially and genetically so in their formation: language, for the realist, arises and operates as a name for the preexisting world. Such a view issues in a bifurcation of linguistic activity into a primary act of reference—concerning what is "real," given "out there" within the prior world waiting to be labeled and denoted—and a subsidiary act of describing, commenting on, and communicating about the objects named. Frege, who never tired of arguing for the opposition between these two linguistic activities—what Mill's earlier realism distinguished as connotation/denotation and he called "sense/reference"—insisted that it was the latter that provided the ground on which mathematics was to be based. And if for technical reasons Frege's ground—the preexisting world of pure logical objects—is no longer tenable and is now replaced by an abstract universe of sets, then his insistence on the priority of reference over sense remains as the linguistic cornerstone of twentieth-century Platonism.

What is wrong with it? Why should one not believe that mathematics is about some ideal timeless world populated by abstract unchanging objects; that these objects exist, in all their attributes, independent of any language used to describe them or human consciousness to apprehend them; and that what a theorem expresses is objectively the case, an eternally true description of a specific and determinate state of affairs about these objects?

One response might be to question immediately the semiotic coherence of a *pre*linguistic referent: "Every attempt to establish what the referent of a sign is forces us to define the referent in terms of an abstract entity which moreover is only a cultural convention" (Eco 1976, 66). If such is the case, then language—in the form of cultural mediation—is inextricable from the process of referring. This will mean that the supposedly distinct and opposing categories of reference and sense interpenetrate each other and that the object referred to can neither be separated from nor antedate the descriptions given of it. Such a referent will be a social historical construct; and, notwithstanding the fact that it might present itself as abstract, cognitively universal, presemiotic (as is the case for mathematical objects), it will be no more timeless, spaceless, or subjectless than any other social artifact. On this view mathematical Platonism never gets off the ground, and Frege's claim that mathematical assertions are objectively true about eternal "objects" dissolves into a psychologistic opposite that he would have abhorred: mathematics makes subjective assertions—dubitable and subject to revision—about entities that are time-bound and culturally loaded.

Another response to Platonism's reliance on such abstract referents—one which is epistemological rather than purely semiotic, but which in the end leads to the same difficulty—lies in the questions "How can one come to *know* anything about objects that exist outside space and time?" and "What possible causal chain could there be linking such entities to temporally and spatially situated human knowers?" If knowledge is thought of as some form of justified belief, then the question repeats itself on the level of validation: What manner of conviction and persuasion is there that will connect the Platonist mathematician to this ideal and inaccessible realm of objects? Plato's answer—that the world of human knowers is a shadow of the eternal ideal world of pure form, so that by examining how what can be perceived partakes of and mimics the ideal, one arrives at knowledge of the eternal—succeeds only in recycling the question through the metaphysical obscurities of how concrete and palpable particulars are supposed to partake of and be shadows of abstract universals. How does Frege manage to deal with the problem?

The short answer is that he doesn't. Consider the distinctions behind Frege's insistence that "the thought we express by the Pythagorean theorem is surely timeless, eternal, unchangeable" (1967, 37). Sentences express thoughts. A thought is always the sense of some indicative sentence; it is "something for which the question of truth arises" and so cannot be material, cannot belong to the "outer world" of perceptible things which exists independently of truth. But neither do thoughts belong to the "inner world," the world of sense impressions, creations of the imagination, sensations, feelings, and wishes. All these Frege calls "ideas." Ideas are experienced, they need an experient, a particular person to have them whom Frege calls their "bearer"; as individual experiences every idea has one and only one bearer. It follows that if thoughts are neither inner ideas nor outer things, then

> A third realm must be recognized. What belongs to this corresponds to ideas, in that it cannot be perceived by the senses, but with things, in that it needs no bearer to the contents of whose consciousness to belong. Thus the thought, for example, which we expressed in the Pythagorean theorem is timelessly true, true independently of whether anyone takes it to be true. It needs no bearer. It is not true for the first time when it is discovered, but is like a planet which, already before anyone has seen it, has been in interaction with other planets. (Frege 1967, 29)

The crucial question, however, remains: What is our relation to this noninner, nonouter realm of planetary thoughts, and how is it realized? Frege suggests that we talk in terms of seeing things in the outer world, having ideas in the inner world, and thinking or *apprehending* thoughts in this third world; and that in apprehending a thought we do not create it but come to stand "in a certain relation . . . to what already existed before." Now Frege admits that while "apprehend" is a metaphor, unavoidable in the circumstances, it is not to be given any subjectivist reading, any interpretation that would reduce mathematical thought to a psychologism of ideas: "The apprehension of a thought presupposes someone who apprehends it, who thinks it. He is the bearer of the thinking but not of the thought. Although the thought does not belong to the thinker's consciousness yet something in his consciousness must be aimed at that thought. But this should not be confused with the thought itself" (Frege 1967, 35).

Frege gives no idea, explanation, or even hint as to what this "something" might be which allows the subjective, temporally located bearer to "aim" at an objective, changeless thought. Certainly he sees that there is a difficulty in connecting the eternity of the third realm to the time-bound presence of bearers: he exclaims, "And yet: What value could there be for us in the eternally unchangeable which could neither undergo effects nor have effect on us?" His concern, however, is not an epistemological one about human knowing (how we can know thoughts) but a reverse worry about "value" conceived in utilitarian terms (how can thoughts be useful to us who apprehend them). The means by which we manage to apprehend them are left in total mystery.

It does look as if Platonism, if it is going to insist on timeless truth, is incapable of giving a coherent account of knowing, and a fortiori of how mathematical practice comes to create mathematical knowledge.

But we could set aside Platonism's purely philosophical difficulties about knowledge and its aspirations to *eternal* truths (although such is the principal attraction to its adherents) and think semiotically: we could ask whether what Frege wants to understand by thoughts might not be interpreted in terms of the amalgamations of thinking/scribbling we have called *signs*. Thus, can one not replace Frege's double exclusion (thoughts are neither inner subjectivities nor outer materialities) with a double inclusion (signs are both materially based signifiers and mentally structured signifieds), and in this way salvage a certain kind of semiotic sense from his picture of mathematics as a science of objective truths? Of course, such "truths" would have to relate to the activities of a sign-interpreting agency; they would not be descriptions of some nontemporal extrahuman realm of objects, but rather laws—freedoms and limitations—of the mathematical subject. A version of such an anthropological science seems to have occurred to Frege as a way of recognizing the "subject" without at the same time compromising his obsessive rejection of any form of psychologism: "Nothing would be a greater misunderstanding of mathematics than its subordination to psychology. Neither logic nor mathematics has the task of investigating minds and the contents of consciousness whose bearer is a single person. Perhaps their task could be represented rather as the investigation of the mind, of the mind not minds" (Frege 1967, 35).

But in the absence of any willingness to understand that both minds and mind are but different aspects of a single process of semiosis, that both are inseparable from the social and cultural creation of meaning by sign-using subjects, Frege's opposition of mind/minds degenerates into an unexamined Kantianism that explains little (less than Brouwer's intuitionism, for example) about how thoughts—that is, in this suggested reading of him, the signifieds of assertion signs—come to inhabit and be "apprehended" as objective by individual subjective minds.

In fact, any attempt to rescue Platonism from its incoherent attachment to "eternal" objects can only succeed by destroying what is being rescued: the incoherence, as we said earlier, lies not in the supposed eternality of its referents but instead in the less explicit assumption, imposed by its realist conception of language, that they are prelinguistic, presemiotic, precultural. Only by being so could objects—existing, already out there, in advance of language that comes after them—possess "objective" attributes untainted by "subjective" human interference. Frege's antipsychologism and his obsession with eternal truth correspond to his complete acceptance of the two poles of the subjective/objective opposition—an opposition which is the sine qua non of nineteenth-century linguistic realism. If this opposition and the idea of a "subject" it promotes is an illusion, then so too is any recognizable form of Platonism.

That the opposition is an illusion becomes apparent once one recognizes that mathematical signs play a *creative* rather than merely descriptive function in mathematical practice. Those things that are "described"—thoughts, signifieds, notions—and the means by which they are described—scribbles—are mutually constitutive: each causes the presence of the other; so that mathematicians at the same time think their scribbles and scribble their thoughts. Within such a scheme the attribution of "truth" to mathematical assertions becomes questionable and problematic, and with it the Platonist idea that mathematical reasoning and conviction consists of giving assent to deductive strings of truth-preserving inferences.

On the contrary, as the model that we have constructed demonstrates, the structure of mathematical reasoning is more complicated and interesting than any realist interpretation of mathematical language and mathematical "subjectivity" can articulate. Persuasion and the dialectic of thinking/scribbling that embodies it is a tripartite activity: the Person constructs a narrative, the leading principle of an argument, in the meta-Code; this argument or proof takes the form of a thought experiment in the Code; in following the proof the Subject imagines his Agent to perform certain actions and observes the results; on the basis of these results, and in the light of the narrative, the Person is persuaded that the assertion being proved—which is a prediction about the Subject's sign activities—is to be believed.

By reducing the function of mathematical signs to the naming of presemiotic objects, Platonism leaves a hole at precisely the place where the thinking/scribbling dialectic occurs. Put simply, Platonism occludes the Subject by flattening the trichotomy into a crude opposition: Frege's bearer—subjective, changeable, immersed in language, mortal—is the Person, and the Agent—idealized, infinitary—is the source (although he could not say so) of objective eternal "thoughts." And, as has become clear, it is precisely about the middle term, which provides the epistemological link between the two, that Platonism is silent.

If Platonic realism is an illusion, a myth clothed in the language of some supposed scientific "objectivity," that issues from the metaphysical desire for absolute eternal truth rather than from any nontheistic wish to characterize mathematical activity, then it does nonetheless—as the widespread belief in it indicates—answer to some practical need. To the ordinary mathematician, unconcerned about the nature of mathematical signs, the ultimate status of mathematical objects, or the semiotic basis of mathematical persuasion, it provides a simple working philosophy: it lets him get on with the normal scientific business of research by legitimizing the feeling that mathematical language describes entities and their properties that are "out there," waiting independently of mathematicians, to be neither invented nor constructed nor somehow brought into being by human cognition, but rather *discovered* as planets and their orbits are discovered.

It is perfectly possible, however, to accommodate the force of this feeling without being drawn into any elaborate metaphysical apparatus of eternal referents and the like. All that is needed is the very general recognition familiar since Hegel—that human products frequently appear to their producers as strange, unfamiliar, and surprising; that what is created need bear no obvious or transparent markers of its human (social, cultural, historical, psychological) agency, but on the contrary can, and for the most part does, present itself as alien and prior to its creator.

Marx, who was interested in the case where the creative activity was economic and the product was a commodity, saw in this masking of agency a fundamental source of social alienation, whereby the commodity appeared as a magical object, a fetish, separated from and mysterious to its creator; and he understood that in order to be bought and sold commodities *had* to be fetishized, that it was a condition of their existence and exchangeability within capitalism. Capitalism and mathematics are intimately related: mathematics functions as the grammar of technoscientific discourse that every form of capitalism has relied on and initiated. So it would be feasible to read the widespread acceptance of mathematical Platonism in terms of the effects of this intimacy, to relate the exchange of meaning within mathematical languages to the exchange of commodities, to see in the notion of a "timeless, eternal, unchangeable" object the presence of a pure fetishized meaning, and so on; feasible, in other words, to see in the realist account of mathematics an ideological formation serving certain (technoscientific) ends within twentieth-century capitalism.

But it is unnecessary to pursue this reading. Whether one sees realism as a mathematical adjunct of capitalism or as a theistic wish for eternity, the semiotic point is the same: what present-day mathematicians think they are doing—using mathematical language as a transparent medium for describing a world of presemiotic reality—is semiotically alienated from what they are, according to the present account, doing—namely, creating that reality through the very language which claims to "describe" it.

WHAT IS MATHEMATICS "ABOUT"?

To claim, as I have done, that mathematical thought and scribbling enter into each other, that mathematical language creates as well as talks about its worlds of objects, is to urge a thesis antagonistic not only to the present-day version of mathematical Platonism but also to *any* interpretation of mathematical signs, however sanctioned and natural, that insists on the separateness of objects from their descriptions.

Nothing is nearer to mathematical nature than the integers, the progression of those things mathematicians allow to be called the "natural" numbers. And no opposition is more sanctioned and acknowledged as obvious than that between these *numbers* and their names, the *numerals*, which denote and label them. The accepted interpretation of this opposition runs as follows: first (and the priority is vital) there are numbers, abstract entities of some sort whose ultimate nature, mysterious though it might be, is irrelevant for the distinction in hand; then there are numerals—notations, names such as

1 1 1 1 1 1 1 1 1 1

X, 10, and so on—attached to them. According to this interpretation, the idea that numerals might precede numbers, that the order of creation might be reversed or neutralized, would be dismissed as absurd: for did not the integers named by Roman X or Hindu 10 exist before the Romans took up arithmetic or Hindu mathematicians invented the place notation with zero? And does not the normal recognition that X, 10, "ten," "dix," and so forth, name the *same* number require one to accept the priority of that number as the common referent of these names?

Insisting in this way on the prior status of the integers, and with it the posterior status of numerals, is by no means a peculiarity of Fregean realism. Hilbert's formalism, for all its programmatic abolition of meaningful entities, had in practice to accept that the whole numbers are in some sense given at the outset—as indeed does constructivism, either in the sense of Brouwer's intuitionism, where they are a priori constructions in the intuition of time, or in the version urged by Kronecker according to which "God made the integers, all the rest being the work of man."

In the face of such a universal and overwhelming conviction that the integers—whether conceived as eternal Platonic entities, preformal givens, prior intuitions, or divine creations—are *before* us, that they have always been there, that they are not social, cultural, historical artifacts, but rather *natural* objects, it is necessary to be more specific about the semiotic answer to the fundamental question of what (in terms of sign activity) the whole numbers are or might be.

However possible it is for them to be individually instantiated, exemplified, ostensibly indicated in particular, physically present, pluralities such as piles of stones, collections of marks, fingers, and so on, numbers do not arise, nor can they be characterized, as single entities in isolation from one another: they form an ordered sequence, a *progression.* It seems impossible to imagine what it means for "things" to be the elements of this progression except in terms of their production through the process of *counting.* And since counting rests on the repetition of an identical act, any semiotic explanation of the numbers has to start by invoking the familiar pattern of figures

1, 11, 111, 1111, 11111, 111111, 1111111, etc.,

created by iterating the operation of writing down some fixed but arbitrarily agreed upon symbol type. Such a pattern achieves mathematical meaning as soon as the type "1" is interpreted as the signifier of a mathematical sign and the "etc." symbol as a command, an imperative addressed to the mathematician, which instructs him to enact the rule: copy previous inscription then add to it another type. Numbers, then, appear as soon as there is a subject who counts.

As Lorenzen—from an operationalist viewpoint having much in common with the outlook of the present project—puts it: "Anybody who has the capacity of producing such figures can at any time speak of numbers" (1955, 38). With the semiotic model I have proposed, the subject to whom the imperative is addressed is the Subject, while the one who enacts the instruction, the one who is capable of this unlimited written repetition, is his Agent. Between them they create the possibility of a progression of numbers, which is exactly the ordered sequence of signs whose signifiers are "1," "11," and so on.

Seen in this way, numbers are things in *potentia*, theoretical availabilities of sign production, the elementary and irreducible signifying acts that the Subject, the one-who-counts, can imagine his Agent to perform via a sequence of iterated ideal marks whose paradigm is the pattern 1, 11, 111, etc. The meaning that numbers have—what in relation to this pattern they are capable of signifying within assertions—lies in the imperatives and thought experiments that mathematics can devise to prove assertions; that is, can devise to persuade the mathematician that the predictions being asserted about his future encounters with number signs are to be believed.

Thus, the numbers are objects that result—that is, are capable of resulting—from an amalgam of two activities, thinking (imagining actions) and scribbling (making ideal marks), which are inseparable: mathematicians think about marks they themselves have imagined into potential existence. In no sense can numbers be understood to precede the signifiers that bear them; nor can the signifiers occur in advance of the signs (the numbers) whose signifiers they are. Neither has meaning without the other: they are coterminous, cocreative, and cosignificant.

What then, in such a scheme, is the status of numerals? Just this: since it seems possible to imagine pluralities or collections or sets or concatenates of marks only in the presence of notations that "describe" these supposedly prior pluralities, it follows that every system of numerals gives rise to its own progression of numbers. But this seems absurd and counterintuitive. For is it not so that the "numbers" studied by Babylonian, Greek, Roman, and present-day mathematicians, although each of these mathematical cultures presented them through a radically different numeral system, are the *same* numbers? If they are not, then (so the objection would go) how can we even understand, let alone include within current mathematics, theorems about numbers produced by, say, Greek mathematics? The answer is that we do so through a backward appropriation: mathematics is historically cumulative not because both we and Greek mathematicians are talking about the same timeless "number"—which is essentially the numerals-name-numbers view—but because we refuse to mean anything by "number" that does not square with what we take them to have meant by it. Thus, Euclid's theorem "given any prime number one can exhibit a larger one" is not the same as the modern theorem "there exist infinitely many prime numbers" since, apart from any other considerations, the nature of Greek numerals makes it highly unlikely that Greek mathematicians thought in terms of an infinite progression of numbers. That the modern form subsumes the Greek version is the result not of the timelessness of mathematical objects but rather of a historically imposed continuity—an imposition that is by no means explicitly acknowledged, on the contrary presenting itself as the obvious "fact" that mathematics is timeless.

In relation to this issue one can make a more specific claim, one which I have elaborated elsewhere (Rotman 1993b), that the entire modern conception of integers in mathematics was made possible by the system of signifiers provided by the Hindu numerals based on zero; so that it was the introduction of the sign zero—unknown to either classical Greek or Roman mathematicians—into Renaissance mathematics that *created* the present-day infinity of numbers.

Insofar, then, as the subject matter of mathematics is the whole numbers, we can say that its objects—the things which it countenances as existing and which it is said to be "about"—are unactualized possibles, the potential sign productions of a counting subject who operates in the presence of a notational system of signifiers. Such a thesis, though, is by no means restricted to the integers. Once it is accepted that the integers can be characterized in this way, essentially the same sort of analysis is available for numbers in general. The real numbers, for example, exist and are created as signs in the presence of the familiar extension of Hindu numerals—the infinite decimals—which act as their signifiers. Of course, there are complications involved in the idea of signifiers being infinitely long, but from a semiotic point of view the problem they present is no different from that presented by *arbitrarily* long finite signifiers. And moreover, what is true of numbers is in fact true of the entire totality of mathematical objects: they are all signs—thought/scribbles—which arise as the potential activity of a mathematical subject.

Thus, mathematics, characterized here as a discourse whose assertions are predictions about the future activities of its participants, is "about"—insofar as this locution makes sense—itself. The entire discourse refers to, is "true" about, nothing other than its own signs. And since mathematics is entirely a human artifact, the truths it establishes—if such is what they are—are attributes of the mathematical subject: the tripartite agency of Agent/Subject/Person who reads and writes mathematical signs and suffers its persuasions.

But in the end, "truth" seems to be no more than an unhelpful relic of the Platonist obsession with a changeless eternal heaven. The question of whether a mathematical assertion, a prediction, can be said to *be* "true" (or accurate or correct) collapses into a problem about the tense of the verb. A prediction—about some determinate world for which true and false make sense—might in the future be seen to be true, but only *after* what it foretold has come to pass; for only then, and not before, can what was *pre*dicted be dicted. Short of fulfillment, as is the condition of all but trivial mathematical cases, predictions can only be believed to be true. Mathematicians believe because they are persuaded to believe; so that what is salient about mathematical assertions is not their supposed truth about some world that precedes them but instead the inconceivability of persuasively creating a world in which they are denied. Thus, instead of a picture of logic as a form of truth-preserving inference, a semiotics of mathematics would see it as an inconceivability-preserving mode of persuasion—with no mention of "truth" anywhere.

AFTERWORD

The above was written when nobody, as far as I was aware, was interested in constructing a semiotics of mathematics. Certainly, there was nothing in the literature in the late 1980s that could serve as a basis for examining mathematical practice as a business of manipulating written signs; nothing, that is, outside a few fragmentary remarks of Charles Peirce suggesting that mathematical thought was exemplary of the kind of hypothetical thought we all perform all the time to ourselves. What I would have liked was a linguistic/semiotic framework well grounded in natural language that was abstract enough to include the making of meaning in mathematics. This would have provided a way in to the question that preoccupied me then: What is the nature—the genesis, structure, and possibilities—of the mathematical subject, the one who does mathematics? How, in other words, are mathematical symbols connected to the one-who-signs, counts, imagines, writes, proves, calculates? Following on from this would come the question of foundations: Could a semiotics of mathematics that responded to these questions about the mathematical subject add something new to the stale triad of philosophies that dominated the field of mathematical philosophy? The semiotic model I created from Peirce's suggestion was pressed into the service of responding to both of these questions.

Since then, things have changed. The systemic linguistics approach of M. Halliday has provided a framework for natural languages that has been successfully extended both to codes of visual depiction and to mathematical signs. For mathematics, two initiatives have issued from Halliday's work. Jay Lemke (1998) has generalized Halliday's triple of "meaning-making meta-functions" from natural languages into a general scheme organized around three functions—ideational, interpersonal, organizational—that applies to a wide class of semiotic activity including scientific and mathematical languages. Consistent with this, Kay O'Halloran (1996) has produced a model of mathematical symbolism that characterizes it as a form of semiosis she calls "grammatical density"—this being a hybrid of the lexical density of written language and the grammatical intricacy of spoken language detailed by Halliday. Lemke's work also addresses the visual aspects of mathematical signs by introducing a certain binary—topological versus typological—that gives an account of mathematical and scientific reliance on diagrams as opposed to the purely textual productions which seem sufficient for the humanities.

From a very different perspective comes the work of Edwin Coleman, whose examination of mathematics' writing and notational systems insofar as they constitute a special form of textuality is the subject of his interesting and insufficiently appreciated thesis (1988) and subsequent publications (1990, 1992); it is also the subject (but narrowly confined to investigating the purely formal aspects of Venn diagrams) of the thesis of Sun-Joo Shin (1994). Besides these, there have been other initiatives on the critical and philosophical fronts. Recently, Mortensen and Roberts (1997) urged the construction of a semiotically based approach to the foundations of mathematics; and Paul Ernest (1998) adopts a Vygotskian approach to mathematical meaning and is currently adapting the thought-experimental model of agencies presented here to an "epistemic subject" specifically geared to educational situations.

CHAPTER TWO

Making Marks on Paper

Even as a simplifying overview of mathematical linguistic practice, Hilbert's formalism was deeply mistaken: neither mathematics nor metamathematics can be described as the manipulation of "meaningless marks on paper." On the contrary, mathematical marks are filled with and can never be divorced from various kinds of meaning. But his slogan points to a fundamental truth (one reason why it has been so useful); namely, mathematics is a form of graphism, an inscriptional practice based on a system of *writing*. But more than a system, it is a *language* whose symbols, figures, notations, graphs, diagrams, equations, and metadiscursive devices are manipulated according to procedures that depend on and mediate meanings attached to them. As an amalgam of thinking and writing, mathematics rests, then, not on mere marks but on written signs. To say this is to insist on an obvious and simple truth that becomes less transparent as soon as one starts to probe what, in such a context, is intended by the term *writing*.

Everyday usage has "writing" as the inscribing of speech, that is, the recording of spoken words by a system of (usually alphabetic) marks that allow the suitably trained reader to reconstruct the original spoken sounds. The writing of mathematical signs does not fall under such a description: despite numerous understandings to the contrary, mathematical signs do not record or code or transcribe any language prior to themselves. They certainly do not arise as abbreviations or symbolic transcriptions of words in some natural language. If anything, although there's more to it than that, it's quite the opposite: the symbol "√" is not a mathematical sign for "is the square root of," neither is "=" the sign for "is equal to"; rather, these

English locutions are renderings of mathematical notions prior to them, and only after habitual back-and-forth traffic between them can the English phrases be mistaken as the origin of the mathematical signs in question. To appreciate some of the implications of this inversion of natural language terms and mathematical symbols, we need to be more specific about how the ideogrammatic and diagrammatic inscriptions of mathematics differ—conceptually and orthographically—from written words.

In the epilogue to his essay on the development of writing systems, Roy Harris declares:

> It says a great deal about Western culture that the question of the origin of writing could be posed clearly for the first time only after the traditional dogmas about the relationship between speech and writing had been subjected both to the brash counter propaganda of a McLuhan and to the inquisitorial skepticism of a Derrida. But it says even more that the question could not be posed clearly until writing itself had dwindled to microchip dimensions. Only with this . . . did it become obvious that the origin of writing must be linked to the future of writing in ways which bypass speech altogether. (1986, 158)

Harris's intent is programmatic. The passage continues with the injunction not to "replough McLuhan's field, or Derrida's either," but sow them, so as to produce eventually a "history of writing as writing."

Preeminent among dogmas that block such a history is alphabeticism: the insistence that we interpret all writing—understood for the moment as any systematized graphic activity that creates sites of interpretation and facilitates communication and sense making—along the lines of alphabetic writing, as if it were the inscription of prior speech ("prior" in an ontogenetic sense as well as the more immediate sense of speech first uttered and then written down and recorded). Harris's own writings in linguistics as well as Derrida's program of deconstruction, McLuhan's efforts to dramatize the cultural imprisonments of typography, and Walter Ong's long-standing theorization of the orality/writing disjunction in relation to consciousness, among others, have all demonstrated the distorting and reductive effects of the subordination of graphics to phonetics and have made it their business to move beyond this dogma. Whether, as Harris intimates, writing will one day find a speechless characterization of itself is impossible to know, but these displacements of the alphabet's hegemony have already resulted in an open-ended and more complex articulation of the writing/speech couple, especially in relation to human consciousness, than was thinkable before the microchip.

A written symbol long recognized as operating nonalphabetically—even by those deeply and quite unconsciously committed to alphabeticism—is that of number, the familiar and simple other half, as it were, of the alphanumeric keyboard. But, despite this recognition, there has been no sustained attention to mathematical writing even remotely matching the enormous outpouring of analysis, philosophizing, and deconstructive opening up of what those in the humanities have come simply to call "texts."

Why, one might ask, should this be so? Why should the sign system long acknowledged as the paradigm of abstract rational thought and the without-which-nothing of Western technoscience have been so unexamined, let alone analyzed, theorized, or deconstructed, as a mode of writing? One answer might be a second-order or reflexive version of Harris's point about the microchip dwindling of writing, since the very emergence of the microchip is inseparable from the action and character of mathematical writing. Not only would the entire computer revolution have been impossible without mathematics as the enabling conceptual technology (the same could be said in one way or another of all technoscience), but, more crucially, the computer's mathematical lineage and intended application as a calculating/reasoning machine hinges on its autological relation to mathematical practice. Given this autology, mathematics would presumably be the last to reveal itself and declare its origins in writing. (I shall return to this later.)

A quite different and more immediate answer stems from the difficulties put in the way of any proper examination of mathematical writing by the traditional characterizations of mathematics—Platonic realism or various intuitionisms—and by the moves they have legitimated within the mathematical community. Platonism is the contemporary orthodoxy. In its standard version it holds that mathematical objects are mentally apprehensible and yet owe nothing to human culture; they exist, are real, objective, and "out there," yet are without material, empirical, embodied, or sensory dimension. Besides making an enigma out of mathematics' usefulness, this has the consequence of denying or marginalizing to the point of travesty the ways in which mathematical signs are the means by which communication, significance, and semiosis are *brought about.* In other words, the constitutive nature of mathematical writing is invisibilized, mathematical language in general being seen as a neutral and inert medium for describing a given prior reality—such as that of number—to which it is essentially and irremediably posterior.

With intuitionist viewpoints such as those of Brouwer and Husserl, the source of the difficulty is not understood in terms of some *external* metaphysical reality, but rather as the nature of our supposed internal intuition of mathematical objects. In Brouwer's case this is settled at the outset: numbers are nothing other than ideal objects formed within the inner Kantian intuition of time that is the condition for the possibility of our cognition, which leads Brouwer into the quasi-solipsistic position that mathematics is an essentially "languageless activity." With Husserl, whose account of intuition, language, and ideality is a great deal more elaborated than Brouwer's, the end result is nonetheless a complete blindness to the creative and generative role played by mathematical writing. Thus, in "The Origin of Geometry," the central puzzle on which Husserl meditates is "How does geometrical ideality . . . proceed from its primary intrapersonal origin, where it is a structure within the conscious space of the first inventor's soul, to its ideal objectivity?" (1981, 257). It must be said that Husserl doesn't, in this essay or anywhere else, settle his question. One suspects that, indeed, it is incapable of solution. Rather, it is the premise itself that has to be denied: it is the coherence of the idea of primal (semiotically immediated) intuition lodged originally in any individual consciousness that has to be rejected. On the contrary, does not all mathematical intuition—geometrical or otherwise—come into being in relation to mathematical signs, making it both external/intersubjective and internal/private from the start? But to pursue such a line one has to credit writing with more than a capacity to, as Husserl has it, "document," "record," and "awaken" a prior and necessarily prelinguistic mathematical meaning. This is precisely what his whole understanding of language and his picture of the "objectivity" of the ideal prevents him from doing. One consequence of what we might call the documentist view of mathematical writing, whether Husserl's or the standard Platonic version, is that the intricate interplay of imagining and symbolizing, familiar on an everyday basis to mathematicians within their practice, goes unseen.

Nowhere is the documentist understanding of mathematical language more profoundly embraced than in the foundations of mathematics, specifically, in the Platonist program of rigor instituted by twentieth-century mathematical logic. Here the aim has been to show how all of mathematics can be construed as being about sets and, further, can be translated into axiomatic set theory. The procedure is twofold. First, vernacular mathematical usage is made informally rigorous by having all of its terms translated into the language of sets. Second, these informal translations are completely formalized, that is, further translated into an axiomatic system consisting of a Fregean first-order logic supplemented with the extralogical symbol for set membership.

To illustrate, let the vernacular item be the theorem of Euclidean geometry, which asserts that, given any triangle in the plane, one can draw a unique inscribed circle:

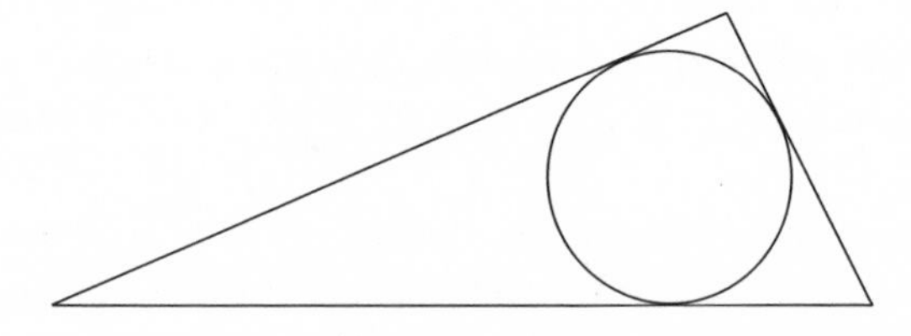

The first translation removes all reference to agency, modality, and physical activity, signaled here in the expression "one can draw." In their place are constructs written in the timeless and agentless language of sets. Thus, first the plane is identified with the set of all ordered pairs (x, y) of real numbers and a line and circle are translated into certain unambiguously determined subsets of these ordered pairs through their standard Cartesian equations; then "triangle" is rendered as a triple of nonparallel lines; and "inscribed" is given in terms of a "tangent," which is explicated as a line intersecting that which it "touches" in exactly one point. The second translation converts the asserted relationship between these abstracted but still visualizable sets into the de-physicalized and de-contextualized logico-syntactical form known as the first-order language of set theory. This will employ no linguistic resources whatsoever other than variables ranging over real numbers, the membership relation between sets, the signs for an ordered pair and for equality, the quantifiers "for all x" and "there exists x," and the sentential connectives "or," "and," "not," and so on.

Once such a double translation of mathematics is effected, metamathematics becomes possible, since one can arrive at results about the whole of vernacular mathematics by proving theorems about the formal (that is, mathematized) axiomatic system. The outcome has been an influential and rich corpus of metamathematical theorems (associated with Skolem, Gödel, Turing, and Cohen, among many, many others). Philosophically, however, the original purpose of the whole foundational enterprise was to illuminate the nature of mathematics by explaining the emergence of paradox, clarifying the horizons of mathematical reasoning, and revealing the status of mathematical objects. In relation to these aims the set theoretization of mathematics and the technical results of metamathematics are unimpressive: not only have they resulted in what is generally acknowledged to be a barren and uninformative philosophy of mathematics but also (and not independently) they have failed to shed any light whatsoever on mathematics as a signifying practice. We need, then, to explain the reason for this impoverishment.

Earlier, I spelled out a semiotic account of mathematics, particularly the interplay of writing and thinking, by developing a model of mathematical activity—what it means to make the signs and think the thoughts of mathematics—intended to be recognizable to its practitioners. The model is based on the semiotics of Charles Sanders Peirce, which grew out of his program of pragmaticism, the general insistence that "the meaning and essence of every conception lies in the application that is to be made of it" (1958, 5:332). He understood signs accordingly in terms of the uses we make of them, a sign being something always involving another—interpreting—sign in a process that leads back eventually to its application in our lives by way of a modification of our habitual responses to the world. We acquire new habits in order to minimize the unexpected and the unforeseen, to defend ourselves "from the angles of hard fact" that reality and brute experience are so adept at providing. Thought, at least in its empirically useful form, thus becomes a kind of mental experimentation, the perpetual imagining and rehearsal of unforeseen circumstances and situations of possible danger. Peirce's notion of habit and his definition of a sign are rich, productive, and capable of much interpretation. They have also been much criticized; his insistence on portraying all instances of reasoning as so many different forms of disaster avoidance is obviously unacceptably limiting. In this connection, Samuel Weber has suggested that Peirce's "attempt to construe thinking and meaning in terms of 'conditional possibility,' and thus to extend controlled laboratory experimentation into a model of thinking in general," should be seen as an articulation of a "phobic mode of behavior," where the fear is that of ambiguity in the form of cognitive oscillation or irresolution, blurring or shifting of boundaries, imprecision, or any departure from the clarity and determinateness of either/or logic (Weber 1987, 30).

Now, it is precisely the elimination of these phobia-inducing features that reigns supreme within mathematics. Unashamedly so: mathematicians would deny that their fears were pathologies, and would, on the contrary, see them as producing what is cognitively and aesthetically attractive about mathematical practice as well as being the source of its utility and transcultural stability. This being so, a model of mathematics utilizing the semiotic insights of Peirce—himself a mathematician—might indeed deliver something recognizable to those who practice mathematics. The procedure is not, however, without risks. There is evidently a self-confirmatory loop at work in the idea of using such a theory to illuminate mathematics, in applying a phobically derived apparatus, as it were, to explicate an unrepentant instance of itself. In relation to this, it is worth remarking that Peirce's contemporary, Ernst Mach, argued for the importance of thought—experimental reasoning to science from a viewpoint quite different from Peirce's semiotics, namely, that of the physicist. Indeed, thought experiments have been central to scientific persuasion and explication from Galileo to the present, figuring decisively in the twentieth century, for example, in the original presentation of relativity theory as well as in the Einstein-Bohr debate about the nature of quantum physics. They have, however, only recently been given the sort of sustained attention they deserve. Doubtless, part of the explanation for this comparative neglect of experimental reasoning lies in the systematizing approach to the philosophy of science that has foregrounded questions of rigor (certitude, epistemological hygiene, formal foundations, exact knowledge, and so on) at the expense of everything else and, in particular, at the expense of any account of the all-important persuasive, rhetorical, and semiotic content of scientific practice.

In any event, the model I propose theorizes mathematical reasoning and persuasion in terms of the performing of thought experiments or waking dreams: one does mathematics by propelling an imago—an idealized version of oneself that Peirce called the "skeleton self"—around an imagined landscape of signs. This model depicts mathematics, by which I mean here the everyday doing of mathematics, as a certain kind of traffic with symbols, a written *discourse* in other words, as follows: all mathematical activity takes place in relation to three interlinked arenas—Code, meta-Code, Virtual Code. These represent three complementary facets of mathematical discourse; each is associated with a semiotically defined abstraction, or linguistic actor—Subject, Person, Agent, respectively—that "speaks," or uses, it. The diagram on p. 52 summarizes these actors and the arenas in relation to which they operate as an interlinked triad.

The Code embraces the total of all rigorous sign practices—defining, proving, notating, and manipulating symbols—sanctioned by the mathematical community. The Code's user, the one-who-speaks it, is the mathematical *Subject*. The Subject is the agency who reads/writes mathematical texts and has access to all and only those linguistic means allowed by the Code. The meta-Code is the entire matrix of unrigorous mathematical procedures normally thought of

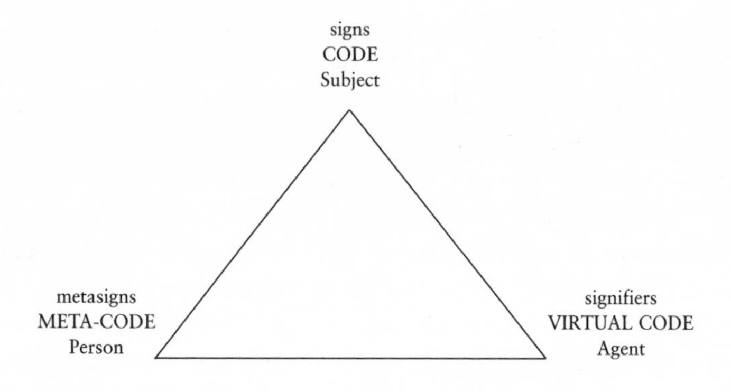

as preparatory and epiphenomenal to the real—proper, rigorous—business of doing mathematics. Included in the meta-Code's resources would be the stories, motives, pictures, diagrams, and other so-called heuristics that introduce, explain, naturalize, legitimate, clarify, and furnish the point of the notations and logical moves that control the operations of the Code. The one-who-speaks the meta-Code, the *Person*, is envisaged as being immersed in natural language, with access to its metasigns and constituted thereby as a self-conscious subjectivity in history and culture. Lastly, the Virtual Code is understood as the domain of all legitimately imaginable operations, that is, as signifying possibilities available to an idealization of the Subject. This idealization, the one-who-executes these activities, the *Agent*, is envisaged as a surrogate or proxy of the Subject, imagined into being precisely in order to act on the purely formal, mechanically specifiable correlates—signifiers—of what for the Subject is meaningful via signs. In unison, these three agencies make up what we ordinarily call "the mathematician."

Mathematical reasoning is thus an irreducibly tripartite activity in which the Person (Dreamer awake) observes the Subject (Dreamer) imagining a proxy—the Agent (Imago)—of him- or herself, and, on the basis of the likeness between Subject and Agent, comes to be persuaded that what the Agent experiences is what the Subject *would* experience were he or she to carry out the unidealized versions of the activities in question. We might observe in passing that the three-way process at work here is the logico-mathematical correlate of a more general and originating triangularity inherent to the usual divisions invoked to articulate self-consciousness: the self-as-object instantiated here by the Agent, the self-as-subject by the Subject, and the sociocultural other, through which any such circuit of selves passes, by the Person.

Two features of this way of understanding mathematical activity are relevant here: First, mathematical assertions are to be seen, as Peirce insisted, as foretellings, predictions made by the Person about the Subject's future engagement with signs, with the result that the process of persuasion is impossible to comprehend if the role of the Person as observer of the Subject/Agent relation is omitted. Second, mathematical thinking and writing are folded into each other and are inseparable not only in an obvious practical sense but also theoretically, in relation to the cognitive possibilities that are mathematically available. This is because the Agent's activities exist and make narrative logical sense (for the Subject) only through the Subject's manipulation of signs in the Code.

The second feature of this model, the thinking/writing nexus, will occupy us below. On the first, however, observe that there is an evident relation between the triad of Code / meta-Code / Virtual Code here and the three levels—rigorous/vernacular/formal—of the Platonistic reduction illustrated above. Indeed, in terms of external attributes, the difference drawn by the mathematical community between unrigorous/vernacular and rigorous/set-theoretical mathematics seems to map onto that between the meta-Code and the Code. This is indeed the case, but the status of this difference is here inverted and displaced. On the present account, belief in the validity of reasoning, or acquiescence in proof, takes place only when the Person is persuaded, a process that hinges on a judgment—available only to the Person—that the likeness between the Subject and the Agent justifies replacing the former with the latter. In terms of our example, the proof of the theorem in question lies in the relationship between the Person, who can draw a triangle and see it as a drawn triangle; the Subject, who can replace this triangle with a set-theoretical description; and the Agent, who can act upon an imagined version of this triangle. By removing all reference to agency, the Platonistic account renders this triple relation invisible. Put differently, in the absence of the Person's role, no explication of conviction—without which proofs are not proofs—can be given. Instead, all one can say about a supposed proof is that its steps, as performed by the "mathematician," are logically correct, a truncated and wholly unilluminating description of mathematical reasoning found and uncritically repeated in most contemporary mathematical and philosophical accounts. But once the Person is acknowledged as vital to the mathematical activities of making and proving assertions, it becomes impossible to see the meta-Code as a supplement to the Code, as a domain of mere psychological/motivational affect, to be jettisoned as soon as the real, proper, rigorous mathematics of the Code has been formulated.

If we grant this, then we are faced with a crucial difference operating within what are normally and uncritically called mathematical "symbols," a difference whose status not only is misperceived within the contemporary Platonistic program of rigor, but, beyond this, is treated within that program to a reductive alphabeticization. The first split in the following diagram, the division of writing into the alpha and the numeric, is simply the standard recognition of the nonalphabetic character of numeric, that is, mathematical, writing.

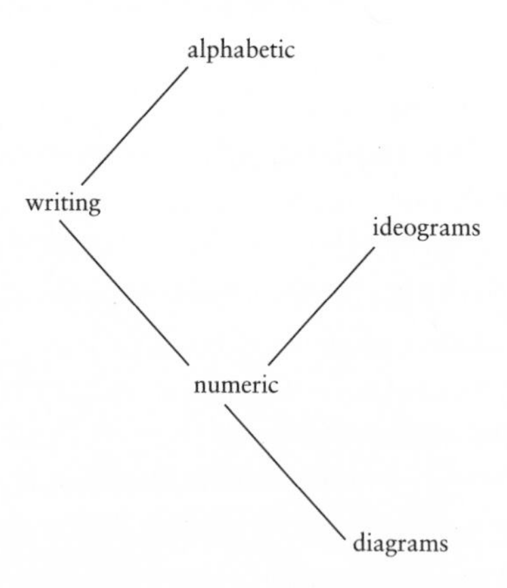

The failure to make this distinction, or rather making it but subsuming all writing under the rubric of the alphabet, is merely an instance of what we earlier called the *alphabetic dogma*. The point of the latter diagram is to indicate the replication of this split, via a transposed version of itself, within the contemporary Platonistic understanding of mathematics. Thus, on the one hand, there are ideograms, such as “+,” “X,” “I,” “2,” “3,” “=,” “>,” “ . . . ,” “sinz,” “logz,” and so on, whose introduction and interaction are controlled by rigorously specified rules and syntactic conventions. On the other hand, there are diagrams, visually presented semiotic devices, such as the familiar lines, axes, points, circles, and triangles, as well as all manner of figures, markers, graphs, charts, commuting circuits, and iconically intended shapes. On the orthodox view, the difference is akin to that between the rigorously literal, clear, and unambiguous ideograms and the metaphorically unrigorous diagrams. The transposition in question is evident once one puts this ranking of the literal over the metaphorical into play: as soon as one accepts the idea that diagrams, however useful and apparently essential for the actual doing of mathematics, are nonetheless merely figurative and eliminable and that mathematics, in its proper rigorous formulation, has no need of them. Within the Platonist program, this alphabetic prejudice is given a literal manifestation: linear strings of symbols in the form of normalized sequences of variables and logical connectives drawn from a short, preset list determine the resting place for mathematical language in its purest, most rigorously grounded form.

There is a philosophical connection between this transposed alphabeticism and classical ontology. The alphabetic dogma rides on and promotes an essential secondarity. In its original form, this meant the priority of speech to writing, that is, the insistence that writing is the transcription of an always preceding speech (and, taking the dogma further back, that speech is the expression of a prior thought, which in turn is the mirror of a prior realm . . .). Current Platonistic interpretations of mathematical signs replay this secondarity by insisting that signs are always signs *of* or *about* some preexisting domain of objects. Thus, the time-honored distinction between numerals and numbers rests on just such an insistence that numerals are mere notations—names—subsequent and posterior to numbers existing prior to and independent of them. According to this understanding of signs, it’s easy to concede that numerals are historically invented, changeable, contingent, and very much a human product while maintaining a total and well-defended refusal to allow any of these characteristics to apply to numbers. And what goes for numbers goes for all mathematical objects. In short, contemporary Platonism's interpretation of rigor and its ontology—despite appearances, manner of presentation, and declared motive—go hand in glove: both relying on and indeed constituted by the twin poles of an assumed and never-questioned secondarity.

Similar considerations are at work within Husserl's phenomenological project and its problematic of geometrical origins. Only there, the presemiotic—that which is supposed to precede all mathematical language—is not a domain of external Platonic objects subsequently described by mathematical signs but rather a field of intuition. The prior scene is one of "primal intrapersonal intuition" that is somehow—and this is Husserl's insoluble problem—"awakened" and "reactivated" by mathematical writing in order to become available to all men at all times as an objective, unchanging ideality. The problem evaporates as a mere misperception, however, if mathematical writing is seen not as secondary and posterior to a privately engendered intuition but instead as constitutive of and folded into the mathematical meaning attached to such a notion. What was private and intrapersonal is revealed as already intersubjective and public.

Nowhere is this more so than in the case of diagrams. The intrinsic difficulty diagrams pose to Platonistic rigor—their essential difference from abstractly conceived sets and the consequent need to replace them with ideogrammatic representations—results in their elimination in the passage from vernacular to formalized mathematics. And indeed, set-theoretical rewritings of mathematics, notably, Bourbaki's, but in truth almost all contemporary rigorous presentations of the subject, avoid diagrams like the plague. And Husserl, for all his critique of Platonistic metaphysics, is no different: not only are diagrams absent from his discussion of the nature of geometry, a strange omission in itself, but they don't even figure as an important item to thematize.

Why should this be so? Why, from divergent perspectives and aims, should both Platonism and Husserlian phenomenology avoid all figures, pictures, and visual inscriptions in this way? One answer is that diagrams—whether actual figures drawn on the page or their imagined versions—are the work of the body; they are created and maintained as entities and attain significance only in relation to human visual-kinetic presence, only in relation to our experience of the culturally inflected world. As such, they not only introduce the historical contingency inherent to all cultural activity but also, more to the present point, call attention to the materiality of all signs and of the corporeality of those who manipulate them in a way that ideograms—which appear to denote purely "mental" entities—do not. And neither Platonism's belief in timeless transcendental truth nor phenomenology's search for ideal objectivity, both irremediably mentalistic, can survive such an incursion of physicality. In other words, diagrams are inseparable from perception: only on the basis of our encounters with actual figures can we have any cognitive or mathematical relation to their idealized forms. The triangle-as-geometrical-object that Husserl would ignore, or Platonists eliminate from mathematics proper, is not only what makes it possible to think that there could be a purely abstract or formal or mental triangle but also an always available point of return for geometrical abstraction that ensures its never being abstracted out of the frame of mathematical discourse. For Merleau-Ponty, the necessity of this encounter is the essence of a diagram:

> I believe that the triangle has always had, and always will have, angles the sum of which equals two right angles . . . because I have had the experience of a real triangle, and because, as a physical thing, it necessarily has within itself everything it has ever been able, or ever will be able, to display. . . . What I call the essence of the triangle is nothing but this presumption of a completed synthesis, in terms of which we have defined the thing. (1962, 332)

In fact, the triangle and its generalizations constitute geometry as means as well as object of investigation: geometry is a mode of imagining with and about diagrams.

Indeed, one finds recent commentaries on mathematics indicating a recognition of the diagrammatic as opposed to the purely formal nature of mathematical intuition. Thus, Philip Davis urges a reinterpretation of "theorem" which would include visual aspects of mathematical thought occluded by prevailing set-theoretical rigor, and he cites V. I. Arnold's repudiation of the scorn with which the Bourbaki collective proclaims that, unlike earlier mathematical works, *its* thousands of pages contain not a single diagram (Davis 1993).

Let us return to our starting point, the question of writing, or, as Harris puts it, "writing *as writing*." One of the consequences of my model is to open up mathematical writing in a direction familiar to those in the humanities. As soon as it becomes clear that diagrams (and indeed all the semiotic devices and sources of intuition mobilized by the meta-Code) can no longer be thought of as the unrigorous penumbra of proper—Coded—mathematics, as so many ladders to be kicked away once the ascent to pure, perceptionless Platonic form has been realized, then all manner of possibilities can emerge. All we need to do to facilitate them is to accept a revaluation of basic terms. Thus, there is nothing intrinsically wrong with or undesirable about "rigor" in mathematics. Far from it: without rigor, mathematics would vanish; the question is how one interprets its scope and purpose. On the present account, rigor is not an externally enforced program of foundational hygiene, but rather an intrinsic and inescapable demand proceeding from writing: it lies within the rules, conventions, dictates, protocols, and such that control mathematical imagination and transform mathematical intuition into an intersubjective writing/thinking practice. It is in this sense that, for example, Gregory Bateson's tag line that mathematics is a world of "rigorous fantasy" should be read. Likewise, one can grant that the meta-Code/Code division is akin to the metaphor/literal opposition but refuse the pejorative sense that set-theoretical rigor has assigned to the term "metaphor." Of course, there is a price to pay. Discussions of tropes in the humanities have revealed that no simple or final solution to the "problem" of metaphor is possible; there is an always uneliminable reflexivity, since it proves impossible, in fact and in principle, to find a trope-free metalanguage in which to discuss tropes and so to explain metaphor in terms of something nonmetaphorical. For mathematics the price—if such it be—is the end of the foundational ambition, the desire to ground mathematics, once and for all, in something fixed, totally certain, timeless, and prelinguistic. Mathematics is not a building—an edifice of knowledge whose truth and certainty are guaranteed by an ultimate and unshakable support—but a process: an ongoing, open-ended, highly controlled, and specific form of written intersubjectivity.

What, then, are we to make of mathematical diagrams, of their status as writing? How are they to be characterized vis-à-vis mathematical ideograms, on the one hand, and the words of nonmathematical texts, on the other? It would be tempting to invoke Peirce's celebrated trichotomy of signs—symbol, icon, and index—at this point. One could ignore indexicals and regard ideograms as symbols (signs resting on an arbitrary relation between signifier and signified) and diagrams as icons (signs resting on a motivated connection between the two). Although there is truth in such a division, it is a misleading simplification: the ideogram/diagram split maps only with great artificiality onto these two terms of Peirce's triad. In addition, there is a terminological difficulty: Peirce restricted the term "diagram" to one of three kinds of icon (the others being "image" and "metaphor"), which makes his usage too narrow for what we here, and mathematicians generally, call diagrams. The artificiality arises from the fact that the ideogrammatic cannot be separated from the iconic nor the diagrammatic from the indexical. Thus, not only are ideograms often enmeshed in iconic sign use at the level of algebraic schemata but, more crucially, diagrams, although iconic, are also, less obviously, indexical to varying degrees. Indeed, the very fact of their being physically experienced shapes, of their having an operative meaning inseparable from an embodied and therefore situated gesture, will ensure that this is the case.

But this is a very generic source of indexicality, and some diagrams exhibit much stronger instances. Thus, consider the diagram, fundamental to post-Renaissance mathematics, of a coordinate axis which consists of an extended, directed line and an origin denoted by zero.

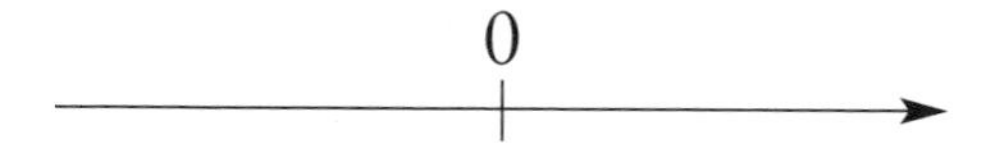

Let's ignore the important but diffuse indexicality brought into play by the idea of directed extension and focus on that of the origin.

Clearly, the function of the ideogram "0" in this diagram is to establish an arbitrarily chosen but fixed and distinguished "here" within the undifferentiated linear continuum. The ideogram marks a "this" with respect to which all positions on the line can be oriented; such is what it means for a sign to function as an origin of coordinates. Indexicality, interpreted in the usual way as a coupling of utterance and physical circumstance, and recognized as present in the use of shifters like "this," "here," "now," and so on, within ordinary language, is thus unambiguously present in our diagram. It does not, however, declare itself as such: its presence is the result of a choice and a determination made in the meta-Code, that is, outside the various uses of the diagram sanctioned within mathematics proper. It is, in other words, the written evidence or trace of an originating act by the Person. Thus, zero, when symbolized by "0," is an ideogrammatic sign for mathematical "nothing" at the same time that it performs a quasi-indexical function within the diagram of a coordinate axis (Rotman 1993b).

It would follow from considerations like these that any investigation of the status of diagrams has to go beyond attempts at classifying them as sign-types and confront the question of their *necessity*. Why does one need them? What essential function—if any—do they serve? Could one do without them? Any answer depends, it seems, on who "one" is: mathematicians and scientists use them as abundantly and with as much abandon as those in the humanities avoid them. In fact, diagrams of any kind are so rare in the texts produced by historians, philosophers, and literary theorists, among others, that any instance sticks out like a sore thumb. An immediate response is to find this avoidance of visual devices totally unsurprising. Would not their embrace be stigmatized as scientism? Indeed, isn't the refusal to use figures, arrows, vectors, and so forth, as modes of explication part of the very basis on which the humanities define themselves as different from the technosciences? Why should texts committed—on whatever grounds—to communicating through words and not primarily interested in the sort of subject matter that lends itself to schematic visual representation make use of it? But this only pushes the question a little further: What allows this prior commitment to words to be so self-sufficient, and what determines that certain topics or subjects, but not others, should lend themselves so readily to diagrammatic commentary and exegesis? Moreover, this separation by content isn't very convincing: philosophers are no less interested in space, time, and physical process than scientists; literary theorists occupy themselves quite as intensely as mathematicians with questions of pattern, analogy, opposition, and structure. Furthermore, whatever its value, such a response gives us no handle on the exceptions, the rare recourse to diagrams, that do occur in humanities texts.

To take a single example, how should we respond to the fact that in Husserl's entire oeuvre there is but one diagram (in his exegesis of temporality), a diagram that, interestingly enough, few commentators seem to make any satisfactory sense of? Are we to think that Husserl, trained as a mathematician, nodded—momentarily slipping from the philosophical into a more mathematical idiom? Or, unable to convey what he meant through words alone, did he resort, reluctantly perhaps but inescapably, to a picture? If the latter, then this fact—the possible inexplicability in words of his account of time—would surely be of interest in any overall analysis of Husserl's philosophical ideas. It would, after all, be an admission—highly significant in the present context—that the humanities' restriction to pictureless texts may be a warding off of uncongenial means of expression rather than any natural or intrinsic self-sufficiency in the face of its subject matter. But then would not this denial, or at any rate avoidance, of diagrams result in texts that were never free (at least never demonstrably so) of a willful inadequacy to their chosen exegetical and interpretive tasks; texts whose wordy opacity, hyperelaboration, and frequent straining of written expression to the edge of sense were the reciprocal cost of this very avoidance?

What goes for diagrams goes (with one exception) for ideograms: their absence is as graphically obvious as that of diagrams; the same texts in the humanities that avoid one avoid the other. And the result is an adherence to texts written wholly within the typographical medium of the alphabet. The exception is, of course, the writing of numbers: nobody, it seems, is prepared to dump the system that writes 7,654,321 in place of seven million, six hundred fifty-four thousand, three hundred twenty-one; the unwieldy prolixity here is too obvious to ignore. But why stop at numbers? Mathematics has many other ideograms and systems of writing—some of extraordinary richness and subtlety—besides the number notation based on 0. What holds philosophers and textual theorists back? Although it doesn't answer this question, we can observe that the place-notation writing of numbers is in a sense a minimal departure from alphabetic typography: an ideogram like 7,654,321 being akin to a word spelled from the "alphabet" 0,1,2,3,4,5,6,7,8,9 of "letters," where to secure the analogy one would have to map the mathematical letter "0" onto something like a hyphen denoting the principled absence of any of the other given letters.

I alluded earlier to Weber's characterization of Peirce's semiotics as founded on a fear of ambiguity and the like. It's hard to resist seeing a reverse phobia in operation here: a recoil from ideograms (and, of course, diagrams) in the face of their potential to disrupt the familiar authority of the alphabetic text, an authority not captured but certainly anchored in writing's interpretability as the inscription of real or realizable speech. The apprehension and anxiety in the face of mathematical grams, which appear here in the form of writing as such—not as a recording of something prior to itself—are that they will always lead outside the arena of the speakable; one cannot, after all, say a triangle. If this is so, then the issue becomes the general relation among the thinkable, the writable, and the sayable, that is, what and how we imagine through different kinds of sign manipulations, and the question of their mutual translatability. In the case of mathematics, writing and thinking are cocreative and, outside the purposes of analysis and the like, impossible to separate.

Transferring the import of this from mathematics to spoken language allows one to see that speech, no less than mathematics, misunderstands its relation to the thinkable if it attempts a separation between the two into prior substance and posterior re-presentation; if, in other words, the form of an always re-presentational alphabetic writing is the medium through which speech articulates how and what it is. By withdrawing from the gram in this way, alphabetic writing achieves the closure of a false completeness, a self-sufficiency in which the fear of mathematical signs that motivates it is rendered as invisibly as the grams themselves. The idea of invisibility here, however, needs qualifying. Derrida's texts, for example, although written within and confined to a pure, diagramless and ideogramless format, nevertheless subvert the resulting alphabetic format and its automatic interpretation in terms of a vocalizable text through the use of various devices: thus a double text such as "Glas," which cannot be the inscription of any single or indeed dialogized speech, and his use of a neologism such as *différance*, which depends on and performs its meaning by being written and not said. But, all this notwithstanding, any attempt to pursue Harris's notion of a speechless link between the origin and the future of writing could hardly avoid facing the question of the meaning and use of diagrams. Certainly, Harris himself is alive to the importance and dangers of diagrams, as is evident from his witty taking apart of the particular diagram—a circuit of two heads speaking and hearing each other's thoughts—used by Saussure to illustrate his model of speech (Harris 1987).

But perhaps such a formulation, although it points in the right direction, is already—in light of contemporary developments—becoming inadequate. Might not the very seeing of mathematics in terms of a writing/thinking couple have become possible because writing is now—postmicrochip—no longer what it was? I suggested above that the reflexivity of the relation between computing and mathematics—whereby the computer, having issued from mathematics, impinges on and ultimately transforms its originating matrix—might be the crux of the explanation for our late recognition of mathematics' status as writing.

To open up the point, I turn to a phenomenon within the ongoing microchip revolution, namely, the creation and implementation of what has come to be known as virtual reality. Although this might seem remote from the nature and practice of mathematics and from the issues that have so far concerned us, it is not, I hasten to add, *that* remote. In addition to many implicit connections to mathematical ideas and mathematically inspired syntax via computer programming, there are explicit links: thus, for example, Michael Benedikt, in his introductory survey of the historical and conceptual context of virtual reality, includes mathematics and its notations as an important thread running through the concept (1992).

An extrapolation of current practices more heralded, projected, and promised than as yet effectively realized, virtual reality comprises a range of effects and projects in which certain themes and practices recur. Thus, one always starts from the given world—the shared, intersubjective, everyday reality each of us inhabits. Within this reality is constructed a subworld, a space of virtual reality that we—or rather certain cyberneticized versions of ourselves—can, in some sense, enter and interact with. The construction of this virtuality, how it is realized—its parameters, horizons, possibilities, and manifestations—varies greatly from case to case. Likewise, what is entailed by a "version" of ourselves, and hence the sense in which "we" can be said to be "in" such virtual arenas, varies, since it will depend on what counts, for the purposes at hand, as physical immersion and interaction and on how these are connected and eventually implemented. In all cases, however, virtual arenas are brought into existence inside computers and are entered and interacted with through appropriate interface devices and prosthetic extensions, such as specially adapted pointers, goggles, gloves, helmets, body sensors, and the like. Perhaps the most familiar example is dipping a single finger into a computer environment via the point-and-click operation of a computer mouse. But a mouse is a very rudimentary interface device, one that gives rise to a minimal interpretation, both in what the internalized finger can achieve as a finger and because a finger is, after all, only a metonym of a body: all current proposals call for more comprehensive prosthetics and richer, more fully integrated modes of interaction with/within these realities once they are entered.

Let's call the self in the world the default or *real-I*; the cyberneticized self we propel around a virtual world, the surrogate or *virtual-I*; and the self mediating between these, as the enabling site and means of their difference, the jacked-in or *goggled-I*. Operating a virtually real environment involves an interplay or a circulation among these three agencies, which ultimately changes the nature of the original, default reality, that is, of what it means to be a real-I inhabiting a/the given world. This circulation and especially its effect on, ultimately its transformation of, the given world motivate a great deal of virtual-reality thinking. To fix the point I'll mention two recent, differently conceived proposals, a social-engineering project and a fantasy extrapolation, both of which explore the possibilities offered by a virtualization of reality. The first, *Mirror Worlds*, is part propaganda, part blueprint for a vast series of public software projects by computer scientist David Gelernter (1992); the second, *Snow Crash*, is a science fiction epic of the near-cybered future by novelist Neal Stephenson (1992).

Mirror Worlds sets itself the task of mapping out, more or less in terms of existing software technology, a way of virtualizing a public entity, such as a hospital or university or city (more ambitiously, an entire country, ultimately the world). Its aim is to create a virtual space, a computer simulation of, say, the city—what Gelernter calls the "agent space"—that each citizen could enter through various interface tools and engage in activities (education, shopping, information gathering, witnessing public events, monitoring and participating in cultural and political activity, meeting other citizens, and so on) in virtual form. The idea is that the results of such virtual-I activities would reflect back on society and effect changes in what it means to be a citizen within a community—to be a real-I—changes in previously unattainable and, given Gelernter's downbeat take on contemporary fragmentation and anomie, sorely needed ways. In *Snow Crash,* Stephenson posits an America whose more computer-savvy denizens can move between a dystopian reality (panoptic surveillance and Mafia-franchised suburban enclaves) and a freely created, utopian computer space, the "metaverse," where their virtual-I's, or "avatars," can access the information net and converse and interact with one another in various virtual ways. Crucial to the plotting and thematics of Stephenson's narrative is the interplay between the inside of the metaverse and the all-too-real outside; the circulation, in other words, of affect and effects between virtual-I's and real-I's as the characters put on and take off their goggles. Although they move in opposite directions—in Stephenson's fantasy the virtual world in the end reflects the intrigue and violence of the real, while for Gelernter the virtual world is precisely the means of eliminating the anomic violence of the contemporary world—they share the idea of opposed worlds separated, joined, and mutually transformed by an interface.

A certain homology between virtual reality and mathematical

thought, each organized around an analogous triad of agencies, should by now be evident. The virtual-I maps onto the mathematical Agent, the real-I onto the Person, and the goggled-I onto the Subject. In accordance with this mapping, both virtual reality and mathematics involve phenomenologically meaningful narratives of propelling a puppet—agent, simulacrum, surrogate, avatar, doppelganger, proxy (Peirce's "skeleton self")—of oneself around a virtual space. Both require a technology that gives real-I's access to this space and that controls the capabilities and characteristics of the skeleton-self agent. In both, this technology is structured and defined in terms of an operator, a figure with very particular and necessary features of its own, distinct from the puppet it controls and from the figure—the Person or real-I—occupying the default reality, able to put on goggles and operate in this way. And both are interactive in a material, embodied sense. In this they differ from the practices made possible by literature, which (like mathematics) conjures invisible proxies and identificatory surrogates of ourselves out of writing, and they differ from the media of theater, film, and TV, where (like virtual reality) proxies are not purely imagined, but instead have a visual presence. The difference arises from the fact that although these media allow, and indeed require of, their recipients/participants an active *interpretive* role, this doesn't and can't extend to any real—materially effective—participation: mathematicians *manipulate* signs; virtual realists *act out* journeys.

Mathematics, then, appears to be not only an enabling technology but also a template and precursor, perhaps the oldest one there is, of the current scenarios of virtual reality. But since something new is enabled here, what then (apart from obvious practical differences) distinguishes them? Surely, a principal and, in the present context, quite crucial difference lies in the instrumental means available to the operator-participants: the mathematical Subject's reliance on the writing technologies of ink-and-chalk inscriptions versus the prosthetic extensions available to the virtual-reality operator. Therefore, what separates them is the degree of palpability they facilitate: the gap between the virtuality of a proxy whose repertoire (in the more ambitious projections) spans the entire sensorimotor range of modalities—ambulatory, auditory, proprioceptive, tactile, kinetic—and the invisible, apparently disembodied Agent of mathematics. The virtual space entered by mathematicians' proxies is, in other words, entirely imagined, and the objects, points, functions, numbers, and so on, in it are without sensible form; percepts in the mind's eye rather than in the real eye necessary for virtual participation. Of course, the journeys that mathematical Agents perform, the narratives that can be told about them, the objects with which they react, and the regularities they encounter are strictly controlled by mathematical signs. Connecting these orders of signification, re-creating the writing/thinking nexus through the interactive manipulation of visible diagrams and ideograms and the imagined, invisible states of affairs they signify and answer to, determines what it means to *do* mathematics.

We are thus led to the question: What if writing is no longer confined to inscriptions on paper and chalkboards but becomes instead the creation of pixel arrangements on a computer screen? Wouldn't such a mutation in the material medium of mathematical writing effect a fundamental shift in what it means to think, and do, mathematics? One has only to bear in mind the changes in consciousness brought into play by the introduction of printing—surely a less radical conceptual and semiotic innovation than the shift from paper to screen—to think that indeed it would. The impact of screen-based visualization techniques on current scientific research and on the status of the theory/experiment opposition, as this has been traditionally formulated in the philosophy and history of science, already seems far-reaching. Thus, although primarily concerned with certain aspects of the recent computerization and mathematization of biology, the conclusion of Tim Lenoir and Christophe Lécuyer's investigation, namely, that "visualization *is* the theory," is suggestive far outside this domain.

New types of mathematics—ways of thinking mathematically—have already come into existence precisely within the field of this mutation. Witness chaos theory and fractal geometry, with their essential reliance on computer-generated images (attractors in phase space, self-similar sets in the complex plane, and so on) that are nothing less than new, previously undrawable kinds of diagrams. And, somewhat differently, witness proofs (the four-color problem, classification of finite groups) that exist only as computer-generated entities. Moreover, there's no reason to suppose that this feedback from computer-created imagery and cognitive representations—in effect, a vector from an abstract, imagination-based technology to a concrete, image-based one—to the conceptual technology of mathematics will stop at the creation of new modes for drawing diagrams and notating arguments. But diagrams, because their meanings and possibilities stem from their genesis as physically drawn, bodily perceived objects, are already quasi-kinematic. In light of this, it is necessary to ask why such a process should be confined to the *visual* mode, to the creation of graphics and imagery, and not extended to the other sense modalities? What is to stop mathematics from appropriating the various computer-created ambulatory, kinesthetic, and tactile features made freely available within the currently proposed schemes for virtual reality? Is it unnatural or deviant to suggest that immersion in a virtually realized mathematical structure—walking around it, listening to it, moving and rearranging its parts, altering its shape, dismantling it, feeling it, and even smelling it, perhaps—be the basis for mathematical proofs? Would not such proofs, by using virtual experience as the basis for persuasion, add to, and go far beyond, the presently accepted practice of manipulating ideograms and diagrams in relation to an always invisible and impalpable structure? The understanding of writing appropriate to this conception of doing mathematics, what we might call *virtual writing*, would thus go beyond the "archewriting" set out by Derrida, since it could no longer be conceived in terms of the "gram" without wrenching that term out of all continuity with itself.

On at least one understanding of the genesis of thinking, nothing could be more natural and less deviant than using structures outside ourselves in order to think mathematics. Thus, according to Merlin Donald's recent account, the principal vector underlying the evolution of cognition and, ultimately, consciousness is the development and utilization of external forms of memory: our neuronal connections and hence our cognitive and imaginative capacities resulting, on this view, from forms of storage and organization outside our heads rather than the reverse (1991).

Evidently, natural or unnatural, such a transformation of mathematical practice would have a revolutionary impact on how we conceptualize mathematics, on what we imagine a mathematical object to be, on what we consider ourselves to be doing when we carry out mathematical investigations and persuade ourselves that certain assertions, certain properties and features of mathematical objects, are to be accepted as "true." Indeed, the very rules and protocols that control what is and isn't mathematically meaningful, what constitutes a "theorem," for example, would undergo a sea change. An assertion would no longer have to be something capturable in a sentencelike piece of—presently conceived—writing but could be a configuration meaningful only within a specifically presented virtual reality. Correspondingly, a proof would no longer have to be an argument organized around a written—as presently conceived—sequence of logically connected symbols but instead could take on the character of an external, empirical verification. Mathematics would thus become what it has long denied being: an experimental subject; one which, although quite different from biology or physics in ways yet to be formulated, would be organized nonetheless around an independently existing, computer-created and -reproduced empirical reality.

This union, or rather, this mutually reactive merging, of mathematics and virtual reality—a coming together of a rudimentary and yet-to-be-consummated technology of manufactured presence with its ancient, highly developed precursor—would take the form of a double-sided process. As we've seen, from outside and independent of any mathematical desiderata, the goal of this technology is to achieve nothing less than the *virtualization of the real*, a process that will engender irreversible changes in what for us constitutes the given world, the default domain of the real-I. From the other side, in relation to a mathematics whose objects and structures have a wholly virtual, nonmaterial existence already, the process appears as the reverse, as effectively a *realization of the virtual*, whereby mathematical objects, by being constructed inside a computer, reveal themselves as materially presented and embodied, a process that will likewise cause irreversible and unexpected changes in the meaning we can attach to the idealized real. One could give a more specific content to this by looking at the most extravagantly virtual concept of contemporary mathematics, that of infinity; a concept so inherently metaphysical and spectral as to be unrealizable—actually or even in principle—within the universe we inhabit. Or so I have argued at length elsewhere (Rotman 1993a).

Of course, these remarks on the joint future of mathematics and virtual reality, although they could be supported by looking at computer-inflected practices within the current mathematical scene, are little more than highly speculative extrapolation here. I have included them in order to get a fresh purchase on the notion of a diagram: by overtly generalizing the notion into a virtually realized (that is, physically presented) mathematical structure, one can see how the question of diagrams is really a question of the body. To exclude diagrams—either deliberately as part of the imposition of rigor in mathematics or less explicitly as an element in a general and unexamined refusal to move beyond alphabetic texts and the linear strings of ideograms that mimic them within mathematics—is to occlude materiality, embodiment, and corporeality, and hence the immersion in history and the social that is both the condition for the possibility of signifying and its (moving) horizon.

CHAPTER 3

How Ideal Are the Reals?

The mathematical continuum consists of all possible points on a line:

$$-3 \quad -2 \quad -1 \quad 0 \quad 1/3 \quad 1 \quad 2 \quad e \quad 3 \quad \pi$$

Assumed as background in introductory algebra, closely encountered by anybody who learns coordinate geometry, more intimately present in calculus, the real continuum is a core mathematical concept and the bedrock on which science since the seventeenth century has constructed its measurement-based engagement with the physical world. Apart from the integers themselves, it's difficult to think of anything in the physical sciences more established and less subject to any kind of doubt or misgiving about its status.

Nonetheless, in the past twenty years there have been doubts as well as outright criticism in physics and computer science of the continuum's hallowed role as the basis for assigning a numerical measure to the universe. What follows is an exploration of this criticism. It is in three parts. First, I look at the problems certain physicists have with the evident mismatch between the deeply infinitistic continuum and the necessarily finite nature of our encounter with physical reality. Next, I focus on the relation between the continuum's infinite structure and what it means to compute; that is, what physics might have to say about the ultimate limits of computation. Finally, in the postscript, there is a firsthand account of my own close encounter with calculus and its vicissitudes or even, dare one say it, its possible imminent demise.

GETTING OFF THE REAL LINE

The integers, Kronecker famously urged, were the work of God. Whatever we might feel about the divine origins of the sequence of whole numbers, the real number line is undoubtedly man-made, the history of its making visible as a quite explicit series of augmentings—rationals, irrationals, infinitesimals, algebraic numbers, transcendentals of the whole numbers. We picture the real line as an infinite aggregate one number for each point—where each number is defined as an infinite sum, that is, an infinite decimal of the form 0.333 . . . for a third, 3.14159 . . . for π, 1.41421 . . . for root 2, and so on. It's not necessary to use a decimal base—any base would do—and Cauchy sequences of rational numbers or Dedekind cuts in the ordered set of rationals rather than base-representations are the preferred rigorous form. In all cases, however, the definition of a single real number rests on an infinite construction.

In order to appreciate this point, here's a two-minute history of the mathematical infinite. It consists of three moments.

Moment one, twenty-five centuries ago, belongs to Zeno, who posed the enigma of physical endlessness: Is the line—which we use to represent extension in space or the passage of time—infinitely divisible into ever smaller bits or do we reach a limit beyond which further subdivision is impossible? Zeno offered various paradoxes—the tortoise and the hare, the arrow that never reaches its target, and so on—showing how each possible answer led to absurdity. They were designed to demonstrate the theological conclusion, promulgated by Zeno's master Parmenides, that all movement and change—any account of which would have to invoke one or the other of Zeno's offered choices—is illusory. Aristotle's way out of Zeno's impasse was to distinguish between two kinds of infinity: the completed or *actual* infinite, which he argued should be avoided as dangerously contradictory, and the *potential* infinite—the coming into being of an ever larger but always finite magnitude, which was safe, rational, and paradox free.

Moment two, the Renaissance development of infinite series that provided the basis for a so-called solution to Zeno's paradoxes. This was intimately connected to the introduction of zero into Western mathematics. Indeed, at the end of the seventeenth century, in a treatise called "The Dime," Simon Stevin, the Dutch mathematician-engineer, blown away by the creative—infinitely proliferative—power of zero inside the Hindu-Arabic decimal notation, suggested extending the method to the infinitely long decimals now so familiar to us. Out of this, a century later, emerged the calculus. This rested on the notion of an infinitesimal magnitude, which flatly ignored Aristotle's interdiction of actual infinity. Inconsistencies—which emboldened Bishop Berkeley to mock infinitesimals as the ghosts of departed quantities—arose and were eventually eliminated at the beginning of the nineteenth century: the actualist formulation of points infinitely close being replaced by a rigorous potentialist one of a "limit" framed in terms of the epsilon-delta formalism.

Moment three belongs to Georg Cantor's reinstitution of the actual infinite onto the mathematical scene in the form of an infinity of infinite sets. Once again there was inconsistency and various paradoxes, this time of self-reference. These produced two divergent responses: constructivism in the hands of the intuitionist Jan Brouwer who called for a complete rewriting of classical logic and rejected any sort of actual or completed infinite process; and axiomatic set theory that built the actual infinite into itself as a postulate and claimed to obviate any inconsistencies through the carefully constructed formal language of quantifiers instituted by Frege. This latter, totally infinitized response—together with the Platonism natural to it—has come to dominate twentieth-century mathematical practice.

Returning to the continuum, we see that it embraces infinity on two levels. On the microlevel each point or real number relies on the infinite sequence of integer coefficients defining its decimal expansion. Cantor called the magnitude of any such sequence a countable infinity and denoted it by aleph-zero. On the macrolevel of the line, or any segment of it, there is the infinity of points, called the power of the continuum and denoted by c. Cantor proved by his famous diagonal argument that c was a larger infinite power than aleph-zero and conjectured that it was the next infinity, that is, that c was the smallest noncountable infinity. Later Gödel in 1941 and finally Paul Cohen in 1963 showed that this conjecture—the continuum hypothesis was impossible to prove or refute: it being as consistent to suppose the hypothesis true—that there was no infinity intermediate between *c* and aleph-zero—as to suppose that on the contrary there are 2 or 3 or 17 or a million or indeed infinitely many infinities between aleph-zero and *c*.

Although the undecidability of the continuum hypothesis provides no comfort to Platonists, for whom every meaningful mathematical assertion has to be either true or false, its effect on the centrality and fundamental importance of the real numbers to (Platonist-conceived) mathematics or the sciences has been practically nil. The very idea of a coordinate system and hence of a curve, a surface, a region, a manifold; the motivating concept of a space—vector, Hilbert, or any kind of topological space; the measurement of time, distance, probability, temperature, velocity, wavelength, concentration, charge, energy—all of these either use or start from **R** and its arithmetical structure. Moreover, the continuum's structure incorporates the "continuous" that physics attributes to motion and which, through calculus, has ensured that the assumptions written into the continuum permeate the whole of science.

Plainly, the real continuum is a many-sided and fertile mathematical tool whose fundamental features—*linearity*, *continuity*, *infinity*—have ruled the mathematics-world interface for some four centuries. So, is there a problem here? Why not another four centuries of successful articulating, coding, theorizing, and decoding of the world via real numbers?

Developments in mathematics, physics, and computer science during the past quarter century—predicated on and impossible to imagine without the emergence of fast and readily available all-purpose computation—suggest that the hegemony of the real-number continuum and so the preeminence of linearity, continuity, and infinity as scientific tools, far from extending uninterrupted into the future, may be over. Which means that dethronement of calculus itself might be at hand (a possibility I address in the postscript below). And if the application of continuum-based mathematics shows signs of mortality, what does this mean about the integers from which the real numbers are constructed: Should we be suspicious of their infinity? We shall come back to this later.

We've all heard how fractals, complex dynamical systems, dissipative structures, deterministic chaos, represent fundamentally new, *nonlinear* forms of mathematics, yielding descriptions of everything from earthquakes and dripping taps to heartbeat rhythms, the distribution of interstellar gas clouds, word frequency in natural languages, and so on. Too large a task to explain here why this explosion of nonlinearity has occurred; how, as the physicist Bruce West puts it, "The synthesis of mathematical analysis and computational techniques for computers has enabled scientists to break the linear chain of logic that had encased their mathematical models" (1985, 193). Instead, a few (very schematic) remarks about some of the links in this chain.

A function (more loosely, an arrangement, a coupling, a dynamic phenomenon) is said to be linear if it acts proportionally: if its output (effect) is proportional to its input (cause). In the coordinate plane the graph of such a function—plotting cause against effect—is a line (hence the name) whose slope is the constant of proportionality. In particular, and of fundamental importance, small changes in input produce small changes in output.

Linearity is the guiding principle of calculus, where the derivative of function f is defined as the limit of the ratio $[f(x + h) - f(x)]/h$ as $h \to 0$. In the limit, as one says, the chord joining the points $f(x + h)$ and $f(x)$ become the tangent at $f(x)$. Thus, geometrically, this means that calculus insists that *in the limit all curves are straight lines.*

Hand in hand with this interpretation of linearity is the principle of *decomposition* whereby a complex process can be split into constituent bits, which can be individually studied and then reassembled so that the behavior of the whole is, in some workable sense, the sum of its parts. The phenomenal success of these principles during the eighteenth and nineteenth centuries, from the mechanics of oscillation by Bernouli and Euler through Fourier's analysis of heat flow to Maxwell's equations for electromagnetism, underpinned what one might call the classical paradigm. According to this, nature is fundamentally decomposable and linear: nonlinearities being seen as troublesome, complicated exceptions, to be ignored or dealt with as perturbations of the underlying linear reality.

Significantly, the great success of linearity in the physical sciences was never matched in the earth and life sciences. Indeed, so widespread are nonlinear systems in nature that they are coming to represent the norm. And it is *linearity* that is seen as exceptional and artificial, a recognition that led one of the early researchers, Robert May, to urge: "The mathematical intuition so developed [basically calculus and linear algebra] ill equips students to confront the bizarre behavior exhibited by the simplest discrete, non-linear system. . . . Yet such non-linear systems are surely the rule, not the exceptions, outside the physical sciences" (quoted in West 1985, 154).

A major feature of this decentering of linearity is the effect on *predictability*. In the classical paradigm—as articulated by Laplace—one can predict the effect, the output of any system, to any degree of accuracy provided one knows the cause, the input, to the same degree of accuracy. For linear systems this is reasonable, since near causes are guaranteed to produce near effects. For nonlinear systems an opposing paradigm—first articulated by Poincaré—operates, since no such guarantee exists or in general can exist: minute variation in input can and does produce unlimited variation in output. Such systems, although deterministic (and frequently elementary and very simple), can exhibit the kind of organized disorder known as chaos, and the system's behavior will depend on the geometry of its trajectory—for example, the kind of attractor operating on it—in phase space.

Observe: despite rejecting linearity on the level of the phenomena it investigates, the mathematics of complex systems, chaos, and dissipative structures is unregeneratedly linear and continuum-based on the metalevel of its own theoretical development: treating phase space, for example, as a fluid and applying classical calculus-based theorems of hydrodynamics to study the trajectories of a system.

Fractals provide a vivid aspect of nonlinearity. Impossible to avoid, the seahorse image of a fractal diagram known as the Mandelbroit set is paraded as an icon of mathematical mystery everywhere from techie T-shirts to a recent cover of the *Proceedings of the Modern Language Association*. A phenomenon surely not unrelated to the ideological/theological agenda of mathematical Platonism: "How," Martin Gardner trumpets, "mathematicians who pretend that mathematical structure is not 'out there,' independent of human minds, can view successive enlargements of the Mandelbroit set and preserve their cultural solipsism is hard to comprehend" (1989). Or, as the subtitle of a just published monograph, "Fractals, Chaos, Power Laws," has it, we are "Minutes away from an Infinite Paradise." But since we're holding off from God and paradise right now, I'll pursue this later. The mathematical idea of fractals is straightforward and has to do with the absence of a characteristic scale. If we magnify a portion of the Mandelbroit set we get an image that mimics without ever duplicating the original set. It is possible to do this repeatedly—the set is said to be *self-similar*. Likewise, if we repeatedly magnify a bit of a coastline, the resulting image will be indistinguishable from the original. Likewise for numerous other curves and time series: so that, for example, unless we're told, we can't spot whether a stock market price graph has a timescale of a year, a month, a day, or an hour.

Such self-similarity negates the linearity requirement of calculus: fractal curves do not get ever nearer to straight lines under magnification; although they may be continuous images of **R**, they are certainly not differentiable. In fact, strictly speaking, fractal curves such as coastlines are not curves at all—if by curve one means a *one*-dimensional subset of space—rather they are d-dimensional objects where $d > 1$ is fractional (hence the name). Thus, fractals call for a discontinuous geometry in which the absolute separation of length/area has been erased and in which the very notion of a *line*, in the classical Euclidean sense, has been reduced to the artificial limiting case of an object with fractal dimension $d = 1$. In other words, fractals are both nonlinear and discrete in a way that represents a radical departure from continuum mathematics. They do not, however, reject infinity in any sense: not only are they are constructed in the classical complex plane but also it's their *infinite* self-similarity—the "fact" that they go on duplicating versions of themselves forever—that is the source of their much-advertised charm.

From a very different direction nonlinearity arises as parallel processing within computational theory. Here the denial of the line is a denial of the ordering of its points, of serial progression, of the strict sequentiality built into the prevailing von Neumann / Turing approach. Instead, one has parallel computation employing a plurality of units processing items simultaneously. In terms of nonlinearity, such processing puts in place a systematic shift of perspective, a move from total to partial ordering, from global sequentiality to locally defined networks or graphs. A move that in a certain sense is related to the bifurcations behind the nonlinearity of chaotic phenomena, leading to the suggestion that multiplicative cascading and not linear, additive accretion is the more appropriate mode of basic numerical description in the life sciences. But I'll not run with that here.

Finally, as an aside on nonlinearity, let me mention the two decades of theorizing in the humanities devoted to dissolving and destabilizing linearity and its metaphors. At a theoretical level the very coherence of a ranked binary (such as explaining figurative in terms of a prior literality, writing as secondary to speech, and so on) has been challenged. The hierarchy, in other words, of a leading term over a supplementary one—the nonmathematical analogue of the "greater than" relation of a linear ordering—has been put into question, replaced by a "logic of the supplement" that repudiates any possibility of granting priority or originating status to *either* member of such a difference. More broadly, a variety of critical discourses have decentered various notions of linearity that figure in linear narrative, linear historical process, linear causation, linear movement from content to representation, and replaced them with models and metaphors articulated in terms of circuits, networks, cycles, systems of economic exchange, and feedback loops.

But, to return to mathematics, what of the other fundamental features of the real number continuum besides linearity? What of continuity and infinity—are these too in the process of being displaced? Here we need to go to the physicists.

Since its inception physics has clung to, if not positively embraced, the classical belief that nature does not jump but moves smoothly—continuously—from one state to another. The belief is deep rooted: even quantum theory, founded on the discontinuous jumps inherent in whole numbers of action quanta, writes its description of motion and change in terms of continuity via Schrödinger's wave equation. (And in doing so produces the notorious collapse of the wave packet version of the measurement problem—but that's a completely other story.)

The negation of continuity is the discrete. Continuity implies but isn't implied by infinitude. Hence discreteness is weaker than finitude. But most physicists are not too bothered to make these distinctions; and since aspects of a practice or method can be made finite in different ways, a variety of interpretations of "discrete" or "finite" physics have been proposed. I'll mention here four quite separate ones that have arisen as part of the emergence of computational technology in the last quarter century that recommend that continuity or infinity have no place whatsoever in physics.

Thus, for example, the emergence of fast computation is an explicit motive for Donald Greenspan's rejection of calculus and his early advocacy of discrete modeling. For Greenspan it is not just continuity but the stronger property of infinity that has to be repudiated, a move he well understands will be seen by mathematicians as irreligious: "To deny the concept of *infinity* is as unmathematical as it is un-American. Yet, it is precisely a form of such mathematical heresy upon which discrete model theory is built" (1973, 1). (Given mathematics' transculturality, one wonders why finitude should any more deny America than Brazil, Japan, or Australia; perhaps it clashes with Manifest Destiny and the insistence "In God We Trust.")

Greenspan's justification for sacrilege emphasizes that not only is infinity without any physical meaning (as many since Hilbert have recognized) but also geometrically it is without a wisp of a connection to material realizability. Moreover, Greenspan maintains, infinity is at best unhelpful—methodologically and computationally—to the study of nonlinear systems, and at worst misleading in that it focuses on irrelevant—artifactual—questions. He goes on to develop discrete, calculus-free descriptions of certain simple but basic nonlinear systems in terms of difference—not differential—equations.

On a much grander scale the physicist John Wheeler, pursuing a philosophically radical revision of physics that would go beyond physics' search for particular laws of nature and explain physical law itself, likewise wants the continuum completely abolished from physics: "Modern mathematical logic denies the existence of the conventional number continuum. Physics can do no other but follow suit. No natural way offers itself to do so except base everything on elementary quantum phenomena, with their information-theoretic yes-no character" (1988, 6).

The appeal to mathematical logic here—that sanctum of unquestioned infinity—will no doubt strike many as odd. What Wheeler is referring to is a somewhat unstable mixture of Brouwer's intuitionism, Gödel's idealist notion that the line is not captured by the real numbers, and Quine's insistence that, like physical objects, the irrational numbers are a convenient myth. But Wheeler makes his point, if not the coherence of its justification, clear enough: "The physical continuum, and with it all the beautiful machinery of physics, is myth, is idealization. Existence, what we call reality, is built on the discrete" (1988, 6). Wheeler goes further. "No continuum" is only one of his rallying cries. "No tower of turtles is another":

> No structure, no plan of organization, no framework of ideas underlaid by another structure or level of ideas, underlaid by yet another level, and yet another, *ad infinitum*, down to bottomless blackness. To endlessness no alternative is evident but a loop, such as: Physics gives rise to observer-participancy; observer-participancy gives rise to information; and information gives rise to physics. (1990, 81)

Clearly, the "loop" that elsewhere Wheeler refers to as a "Meaning Circuit" of information and existence, a circular passage "It from bit" and vice versa, is an epistemological one: it refers to levels of discourse. It is not presented, I should emphasize in view of later developments, as an alternative to an endless series of mathematical actions. Wheeler's refusal of an infinite regress—no turtles all the way down—neither repudiates nor suggests any curtailment of the negative integers—which, in their natural ordering are the simplest model of such a tower-of-turtles regress.

A more classically framed, less philosophically visionary version of discrete quantum mechanics has been put forward by T. D. Lee, one with the specific aim of eliminating an infinitude of data in a bounded region of space and time. Lee observes that within current formulations of relativistic field theory based on operator functions embedded in a space-time continuum, it is possible to "conceive of performing an infinite sequence of observations, at $x_1, x_2, \ldots$, all lying infinitesimally close but with the result of field measurements, say $o(x_1), o(x_2), \ldots$ fluctuating violently" (Lee 1983, 1). To negate this possibility, Lee suggests that "a limit should exist on the density of measurements per unit space-time domain. Such a limit would automatically remove all ultraviolet divergencies" (1983, 1–2).

Lee's working out of this suggestion rests on a nonstandard treatment of time—considering it not as a continuous parameter but as a discrete dynamical variable. I won't try to describe it. Enough for present purposes to note the following: Lee's program rests on the introduction of what he calls a "*fundamental length* or time l (in natural units)," so that, "Given any time interval T, the total number N of discrete points that define the trajectory is given by the integer nearest T/l" (Friedberg and Lee 1983, 218). By using this fundamental length, Lee and his collaborators are able to rewrite the standard action integrals as sums over finitely specified trajectories and then use the least action principle to arrive at discrete equations of motion that coincide with the continuum ones when $l \to 0$.

Note that Lee's rejection of the continuum is partial. True, instead of a continuum of operators there are only finitely many, but as he makes quite clear, each can have a "continuous range of eigenvalues that can be observed. The discreteness appears . . . only in the maximal number N of measurements allowed in any finite space-time volume omega." Thus, for Lee physical reality is as continuous as it ever was, it is only *observations* (as copresent measurements) that are discretized. But even here, the restriction is weaker than it appears. The aim was to remove the ability to perform an infinite number of measurements within an infinitesimal space-time volume—the cause of the ultraviolet divergence. Outside achieving this, no bound is imposed: "Any finite number of observations may be made very close to each other in space and in time" (1983, 220). In short, for Lee the universe is not discrete, and although the mathematics used to theorize it might be finite, it is unbounded, that is, potentially infinite. Furthermore, his metatheory, the language he uses to compare his mechanics with the usual version, rests on an infinite mathematics, since it involves $l \to 0$, and any such limit-to-zero process is intrinsically infinitary.

Quite differently, and long before Wheeler's call for the continuum's abolition or Lee's working out of a discrete physics, Richard Feynman (an early student of Wheeler's) expressed a characteristic worry about the infinitistic consequences of operating with the continuum:

> It always bothers me that, according to the laws as we understand them today, it takes a computing machine an infinite number of logical operations to figure out what goes on in no matter how tiny a region of space, and no matter how tiny a region of time. How can all that be going on in that tiny space? Why should it take an infinite amount of logic to figure out what one tiny piece of space/time is going to do? (1967, 57)

Some fifteen years later, discussing the possibility of computers simulating physics, Feynman was still bothered:

> It is going to be necessary that *everything* that happens in a finite volume of space and time would have to be analyzable with a finite number of logical operations. The present theory of physics is not that way, apparently. It allows space to go down into infinitesimal distances, wavelengths to get infinitely great, terms to be summed in infinite order, and so forth; and therefore, if this proposition [that physics is computer-simulatable] is right, physical law is wrong. (1982, 468)

What Feynman calls the present theory is of course responsible for the infinities that continue, in Penrose's recent phrase, "to plague conventional quantum field theory" (1989, 350). And it is only with the removal of such infinities, via a coherent and convincingly *physical* (non-"dippy" is Feynman's word) mathematics, that physics can claim to *explain*, rather than merely calculate, quantum phenomena.

Neither Wheeler nor Feynman produced a worked-out version of a continuum-free—discrete—physics. In particular, neither explains what he means by "discrete" other than an uncritical reliance on the evident discreteness of the integers. Consequently, neither focused on any possible incoherence arising from a mismatch between the assumptions of their mathematics and the world. This means that, assuming "discrete physics" means for Wheeler and Feynman (but evidently not for Lee) a discrete universe, there is the question of whether the techniques used in such rewritings of physics exclude—or should in principle exclude—the use of nondiscrete methods such as limit procedures and calculus. Whether, in other words, it makes sense to worry about eliminating the continuum from the universe only to reinsert continuity or infinity via the mathematics used to describe such a universe.

There is the question of what is to be meant by "finite" here. When Wheeler asserts that nothing beyond elementary yes/no questions and the integers are needed for physics, does he envisage the set of *all* integers? Or does he have in mind a finite but *arbitrary* long sequence of integers? He doesn't say. Likewise, when Feynman speaks of a "finite number" of logical operations, does he mean boundedly finite, in the sense of an upper limit being given beforehand, or does he intend "finite" in the customary sense of arbitrarily large, that is, potentially infinite? What for physics might be the difference?

To raise such questions is to bring us to a certain convergence between physics and computer science (one that ultimately goes back to the bomb project at Los Alamos but which crystallized as an identifiable movement in the late 1970s); a coming-together seen not only by Rolf Landauer but also by Stephen Wolfram, Tommaso Toffoli, and many others, in the form of a two-way relation. Thus, Wolfram writes, "There is a close correspondence between physical processes and computations. On the one hand, theoretical models describe physical processes by computations that transform initial data according to algorithms representing physical laws. And on the other hand, computers themselves are physical systems, obeying physical laws" (1985, 735). In other words, the entire universe *is* (or in some unidentified sense "corresponds to") a computer. Although extreme, the metaphor here is in line with its predecessors—universe as clock, universe as steam engine—that have equated physical reality to the prevailing concept of a machine. However, there is a difference this time around. Computers, unlike clocks and steam engines, are taken to have, in relation to machines in general, a *universal* capability: any (classically deterministic) mechanical process can—it is argued—be emulated by a computation on a universal Turing machine.

With this in mind, let's return to Feynman and his discrete physics. His (and any would-be simulator's) starting point is universality: "What kind of computer are we going to use to simulate physics? Computer theory has been developed to a point where it realizes that it doesn't make any difference; when you get to a universal computer, it doesn't matter how it's manufactured, how it's actually made. Therefore my question is, Can physics be simulated by a universal computer?" (1982, 467). In order to get off the ground, Feynman assumes, then, the existence of a universal computer. This relies on the property built into the definition of Turing machines that the memory space of any such machine is unbounded, that is, *potentially* infinite. Without invoking this fundamental aspect of such machines no proof of universality can be given, and much of the intuition behind the world-as-computer metaphor evaporates—or at least becomes highly problematic. Thus, although we don't know what Feynman intended by a "finite" number of operations, we do know that the very notion of a Turing machine and hence the whole formalism underpinning the computer simulation of physical reality is irretrievably committed to an unbounded, potentially infinite interpretation of number. Whether—given his deep-felt suspicion of infinities and attachment to the palpable constraints of physical realizability—Feynman would or should have been happy with this sort of discreteness is a question that has to be left open.

Enough, then, of the impact of the computer on physics and physicists' attempts to rewrite the foundations of their subject in the light of it. I want now to reverse the vector. We have touched on the computation of physics: the claim, that is, that any bit of physical reality can be modeled as a computational process. Reciprocally, there is the physics of computation: the fact that computers are themselves a part of the physical world, subject no less than anything else to the laws of physics. What does this imply about the real number continuum? And about the integers of which the real numbers are an elaborate outgrowth?

THE PHYSICS OF COMPUTATION

In the last two decades the question of what computers can in fact and in principle do in our universe, what is termed the ultimate physical limits of computation, has been systematically investigated by various physicists and computer scientists. Preeminent among these are Rolf Landauer and his coworkers who have organized their approach around the insistence that all information associated with computational processes, whatever function it performs, is necessarily physical. One consequence of this physicalism is Landauer's contention that classical continuum mathematics is an inappropriate and indeed misleading tool to study a universe that is in all likelihood discrete and that such mathematics has to be eliminated if a self-consistent theory of the limits in question is to be developed. I accept here Landauer's starting point and his conclusion about the continuum, but I argue that his critique of mathematics is not radical enough: not only continuum mathematics but also the density of the rationals and hence the *infinitude of the natural numbers* have to be questioned if a coherent theorization of ultimate computational limits is to be forthcoming.

In a recent article (Landauer 1991) summarizing the work on ultimate limits, programmatically entitled "Information Is Physical," Landauer averred that the fact "there are no unavoidable energy consumption requirements per step in a computer" (23) in relation to dissipation (contrary to what many had believed) is now well established and called attention instead to other issues that remain problematic concerning the "operating environment" of the universe and, more pointedly, the mathematical formalism physicists use to investigate physical law. In order to pursue these I need to reprise some of the history and context laid out by Landauer.

The question of what computers can and can't do in the universe we inhabit has occupied Landauer for a considerable period; since, in fact, his pioneering paper, Landauer (1961) made a thermodynamic connection between information and dissipation. What he showed was that physically irreversible operations dissipate energy: a computation "requires a minimal heat generation, per machine cycle, typically of the order of kT for each reversible operation" (183). Logical irreversibility, typically the erasure of information, is associated in a direct way with physical irreversibility. So that, for example, in order to destroy one bit of information at least $kT\ln2$ units of energy needs to be dissipated, where T is the absolute temperature and k is Boltzmann's constant.

Any computation of the kind familiar to us (word processing, calculation, pattern generation, and so on) is permeated with irreversible steps. This follows from the fact that the logical basis for these computations is one directional, since the standard AND and OR gates assumed by contemporary programming languages shed information: they have binary input and unary output.

The limitative nature of Landauer's result was the starting point of the work of Charles Bennett. In 1973, Bennett showed that while Landauer's absolute minimum of entropy production was unavoidable, the premise of its applicability—the widespread and universally accepted reliance on AND- and OR-based irreversible logic—could be circumvented. Bennett demonstrated that irreversible logic gates and the consequent erasures of information they engender are not essential to computation. (Of course, as Bennett observes, any computation can be made reversible by saving all the information along the way. But this is unacceptable, since it only transfers the activity of erasure—and consequent entropy production—to the machine's memory tape, which, when it comes to be reused, would have to be wiped clean.)

Instead, Bennett argued that it was necessary to institute a wholesale rewriting of computational logic, what Fredkin and Toffoli (1982) elaborated as "conservative logic," to replace the familiar and accepted one that fails to conserve directionality. Such a rewriting demonstrates how each previously irreversible step—basically an erasure that destroyed information—could in principle be replaced by a much longer, more complex computation that used this information positively in a bidirectional way. He accomplished this by giving an abstract specification of a class of Turing machines that would in theory run on reversible logic, as well as specifying a physical example—the biosynthesis of messenger RNA—of reversible computation. Since then others (Benioff, Toffoli, Feynman, Zurek) have added to the list of ideal reversible machines—of both a classically deterministic and quantum kind—as well as providing specific proposals, such as Fredkin gates, for physically implementing reversible logic.

Bennett's achievement (which as a bonus led to a reworking of the Szilard-Gabor-Bouillon solution to Maxwell's Demon) has been widely appreciated and influential and has extended and thematized the agenda, set originally by Landauer, for what will be called here the *information-is-physical approach*. I want now to take a closer look at how, according to this approach, one is to understand the original quest for the ultimate physical limits of computation.

With computation understood in terms of reversibility and governed by the appropriate conservative logic, the major source of dissipation is removed from the scene, and we are left with the effects of thermal vibration arising from the fact that any computational act takes place at a temperature that, by the third law of thermodynamics, is necessarily nonzero. The outcome of this is the production of noise, which degrades signals and produces errors. Leaving aside the various quantum models of reversible computing that have been put forward, and simplifying the discussion of the classical alternative, two quite different approaches have been proposed to deal with noise.

One, the so-called billiard ball or ballistic model, eliminates noise and hence errors *entirely*. It does this by being composed of idealized parts with perfectly reflecting surfaces. But, because they need to be assembled with perfect precision, such machines are held to be unrealistic (at least in their classical formulation), and since I make a certain appeal here to realizability I shall ignore them in what follows.

The other encompasses a class of machines Bennett calls "Brownian." With these the idea is not to deny or attempt to eliminate the thermal environment but to work, as best as possible, with it:

> Brownian computers allow thermal noise to influence the trajectory so strongly that it becomes a random walk through the entire accessible (low potential-energy) portion of the computer's configuration space. In these computers, a simple assemblage of simple parts determines a low-energy labyrinth isomorphic to the desired computation, through which the system executes its random walk, with a slight drift velocity due to a weak driving force in the direction of forward computation. . . . The drift velocity is proportional to the driving force, and hence the energy dissipated approaches zero only in the limit of zero speed. (Bennett 1982, 905)

For such machines the resistance they encounter is like the viscosity of electricity or hydrodynamics rather than conventional static friction: the slower one computes, the less work needed to overcome resistance, that is, the less the dissipation.

With this solution to the problem of computational noise—described in Toffoli (1982) as "*virtually* nondissipative computation" and summarized by Landauer as the ability "to minimize dissipation to any desired extent" (1991)—we reach the fulfillment of what we might call the internal aspect of the information-is-physical agenda: no unavoidable energy requirements of computing arise from purely energy-theoretic, thermodynamical constraints.

"Are there then," Landauer goes on to ask, "no limits imposed by physics?" The answer is that, of course, there are, namely, the external constraints of what is out there. We are likely, he observes, to be in a finite universe; it's difficult to believe that nature would allow an unlimited memory; there is everything from corrosion, degradation, electromigration to cosmic rays and spilled coffee to worry about; if we deal with errors by redundancy and building more massive machines, then we might more quickly run out of parts, and so on and so forth.

In addition to these external effects, Landauer also sees a more abstract difficulty in the way of understanding the ultimate limits of computation:

> In contrast with the physical situation, mathematics has taught us to think in terms of an unlimited sequence of operations. We have all grown up with the sense of values of the mathematician: "Given any epsilon, there exists an *N*, such that. . . ." We can calculate π to any required number of places. But that requires an unlimited memory unlikely to be available in our real physical universe. Therefore all of classical continuum mathematics, normally invoked in our formulation of the laws of physics, is not really physically executable. The reader may object. Can we not define the real numbers within a formal mathematical postulate system? . . . Undoubtedly we can. But physics demands more than that; it requires us to go beyond a closed formal system and to calculate actual numbers. If we cannot distinguish pi from a terribly close neighbor, then all the differential equations that constitute the laws of physics are only suggestive. (Landauer 1991, 28)

Thus, Landauer's response is to reject classical continuum mathematics. But he observes—with greater acuity than does the majority of physicists—that things aren't that simple. Such a rejection, if not thoroughgoing, fails to be consistent:

> Others have, in a variety of ways, suggested that space and time in the universe are not really described by a continuum and that there

> is some sort of discretization, or some limit on the information associated with a limited range of space and time. Most of these investigators, however, consider that to be a description of the *physical* universe and are still willing to invoke continuum mathematics to describe their picture. (1991, 291)

Who these others are we are not told, but it would not be difficult to find examples of physicists fitting the description. In any event, Landauer's quest for a self-consistent theorization of the physics of computation points to two issues: an ontological one of *external features*—the possibilities and limitations of what is out there, "the construction materials and operating environments available in our actual universe"; and an epistemological one of the *mathematical formalism*—whose idealized character leads to the possibility of a mismatch between it and physical reality. These two issues are not, as I shall indicate, able to be separated.

Of course, worries about a lack of fit between mathematics and physical reality are not new. In the early days of quantum theory Eddington had misgivings about the validity of mathematical formalism for distance scales smaller than 10^{-13} centimeters, and recently Roger Penrose articulated similar doubts. Referring to the property of density—between any two real numbers there is always a third—Penrose observed:

> It is not at all clear that physical distances or times can realistically be said to have this property. If we continue to divide up the physical distance between two points, we would eventually reach scales so small that the very concept of distance, in the ordinary sense, could cease to have meaning. It is anticipated that at the quantum gravity, scale of 10^{20}th of the size of a sub-atomic particle, this would indeed be the case. But to mirror the real numbers, we would have to go to scales indefinitely smaller than this: 10^{200}th, 10^{2000}th, . . . of a particle size, for example. It is not at all clear that such absurdly tiny scales have any physical meaning whatever. A similar remark would hold for correspondingly tiny intervals of time. (1989, 86)

Neither Eddington nor Penrose saw fit to make their doubts into any principled refusal of the continuum as a tool for theorizing physical reality. This is not so with Landauer, as we see, and not so with

John Wheeler either (invoked by Landauer), who, as is well known, has campaigned widely against the use of continuum mathematics within the formulation of physical law. The question is whether Wheeler and Landauer go far enough. After all, the real number continuum is itself only an outwork of the integers, an algebraic-topological completion, in fact, of the rationals. Problematizing the gap between the infinitely fine gradations of real numbers and what seems to be the all-too-finite nature of reality only raises the question of why we are so secure about the endless progression of integers and the rationals based on them. Avoiding the reals—deciding beforehand that continuum mathematics is "not really executable" as Landauer has it and should form no part of our formulation of the laws of physics—although it pushes in the right direction, hardly settles the matter. It leaves open the question: Why do we believe that the mathematics of the integers is "executable"? At first blush this seems a bizarre question. How could there possibly be any doubts about the role of the familiar whole numbers in the formulation of physical law? Before I confront Landauer's program with this, let me try to remove some of the question's strangeness by commenting on Penrose's doubts (which are presumably not far removed from those motivating Landauer and Wheeler). What exercises Penrose about the reals, in relation to their use by physics, is their density, which implies a mirroring of real number intervals by unrealistically small distance and timescales. But one doesn't need the continuum to encounter this problem: density is already fully present as a property of the *rationals*. And the rationals are fully present—by a simple finitary operation—as soon as the integers are given. An examination of the integers' "executability," then, seems inescapable.

Observe that neither Landauer, Wheeler, nor any of their coworkers has called for a mathematics that would put the integers into question or suggested that we treat them with the kind of suspicion reserved for the reals. The question arises whether the mathematics on which Landauer's program is based does not make for an incoherence parallel to the one he diagnoses but for *the endless progression of integers rather than the continuum*. Thus, might it not be the case that we have not only continuum mathematics illegitimately describing a discrete universe but also infinitistic mathematics equally illegitimately describing a finite world? If such is the case, then the resulting incoherence would undoubtedly prevent the attainment of the self-consistent theory Landauer calls for. What guarantee is there that this doesn't happen, that there is no mismatch or divergence between the information-is-physical program for examining the ultimate limits of computation and the mathematical apparatus it uses to articulate and carry out this program?

To answer this we need to return to the internal portion of the program that seems (at least to its adherents) settled and unproblematic, namely, the lack of any unavoidable energy requirements per computational step, summarized under the slogan "virtually dissipation-free computing." What, we can ask, does *virtually* mean? Equivalently, what is involved in claiming that—in Landauer's words—the "effects of noise can be offset to *any required degree*"? Or again, what are the mathematical assumptions lying behind the assertion that the dissipation approaches zero "in *the limit of zero speed*"? These expressions, and others, such as "arbitrarily small," "as close to zero as one pleases," and so on, are all translations of the same bit of mathematics, namely, the fundamental idea of a limit. As is well known, this idea assumes the prior existence of the endless progression of integers. But "assume" is too weak a word here: the expression "limit to zero," indeed the entire limit formalism as this is interpreted within contemporary mathematics, is without meaning, cannot be said or made intelligible, except in relation to this unlimited progression. Arbitrarily small rationals and arbitrarily large integers go together.

The employment of the limit formalism raises, then, the obvious parallel to Penrose's question: What is the *physical* meaning of computational speed being *arbitrarily close to zero*? How small a computational distance could be traversed in a unit of time? Or how long would one—*could* one in our universe—wait for the next step of a computation?

Such questions are not idle philosophical speculations. Nor are they unnatural in the context of physics' examination of the status of its own limits. Nor can they be dismissed out of hand. They point to a fundamental difficulty of physics' deployment of mathematical objects that are nameable with ridiculous ease—10^{-100} the diameter of an atom or 10^{-1000} second, and so on—but which lack a smidgen of connection to any existing, projected, or even currently imaginable physical theory.

The limit operation $x \to 0$ is problematic only if the numbers being endless is problematic. So the question we need to ask is: In what sense do the whole numbers go on forever? The standard response runs: if they didn't, then there would be a largest number N; and since it's always possible to add one and get the number $N+1$, there can be no largest number; hence the integers are endless. To which there is an obvious (but all too avoidable) question: *Who*—or what—can always add one more? Or, putting it in terms of the integers, rather than in terms of agency, *where* did the number $N+1$ come from? Where, in general, *do* the numbers come from?

Numbers are inseparable from counting. So the question becomes: Who or what is counting? If you're a Platonist about mathematics, as most scientists are, then that's a dumb question. Numbers, like points, lines, functions, limits, circles, spaces, and every other kind of mathematical object, are just "out there." Timeless and spaceless and originless. They don't answer to a "where?" a "who?" or a "what?" They exist and have always existed: uncreated, ideal, transcendental entities; in this perspective, counting is merely a progression through an already existing infinite sequence of objects. Now Landauer is anything but a Platonist, and this sort of metaphysical realism is clearly unacceptable to the information-is-physical proponents. They, on the contrary, have to address the physicality of endless—that is, arbitrarily long—counting. Moreover, for them counting must, in some sense, boil down to computing. In other words, their characterization of classical, nonquantum, counting would be in terms of a deterministic computing device and would run something like the following.

One has a machine *MU* which, although it is strictly not necessary for work we shall assign to it, can for the sake of simplicity be assumed after the work of Bennett to operate reversibly. The typical step executed by *MU* produces $j+1$ as output when presented with an integer j as input. More precisely, *MU* produces a coding of $j+1$, say in binary notation (or decimal, or, more generally, r-ary positional notation) when given a corresponding coding of j. If we start

MU off with an input of zero and let it run, then we would want to say—and we could check its output whenever we liked—that it was counting. Let us look briefly at the thermodynamic operation of *MU*.

$$101011111 \rightarrow MU \rightarrow 101100000$$

We can write down a simple energy equation for *MU*. Thus, let k be Boltzmann's constant, and suppose *MU* counts up to N operating at speed V and temperature T. Then the entropy term—the energy dissipated at each step—will be $akTV$. Besides dissipation there is energy required to do computational work. First, storage: it is known (Bennett 1984) that energy of "at least kT must be tied up in the physical representation of a bit during its transmission or storage." This means that the energy needed to maintain integrity of the input j during the jth step will be proportional to $kT\log(j)$. Second, processing: as well as maintaining integrity during storage and transmission, *MU* also computes, processing the representation by adding 1 at each step to obtain the new representation (e.g., changing three hundred fifty-one to three hundred fifty-two as illustrated); the energy requirement for this is on average proportional to kT for each step. Taking a summation up to N gives the total energy E_N used by *MU* to count up to N as

$$E_N = kT(aNV + b\log(N!) + cN),$$

for suitable constants a, b, c. Note that the Nth step, i.e., the energy required to go from N to $N+1$, is

$$E_{N+1} - E_N = kT(aV + b\log N + c),$$

which is an increasing and unbounded function of N.

We see from this calculation that for large enough N the energy required by *MU* to count one more time, that is to go from N to $N+1$, is greater than any preassigned quantity. Moreover, if in order to make the dissipation arbitrarily small we make the velocity of *MU* arbitrarily close to 0, then the action involved in counting up to N, for any N, becomes arbitrarily large. But it is by no means evident, in fact it's quite obscure, to imagine how the universe would permit such phenomenona; would permit, that is, greater than any preassigned quantity of energy or quantity of action to be available or take place within the region of space-time enveloping some specific, actualizable machine *MU*.

Matters seem locked into a circle here. We use the integers as part of our mathematical formalism to describe and theorize computing in terms of physical quantities such as energy and action, and we use computing as the model and instantiation of counting from which we derive the integers. Is there a way out of the loop? Within the present context there seems not to be, since the sort of scientific investigation implied by an examination of the ultimate limits of computation seems to require that neither computing nor counting be specifiable except in terms of each other. Of course, one could build and run a "computer" without theorizing the process via mathematics (after all, this is what evolution did to produce our brains). And one could, as I have explained elsewhere (Rotman 1993a), understand the process of counting and the whole numbers associated with it as a form of repetitive semiotic activity that didn't mention computers. But outside an evolutionary account of computation or a semiotic understanding of number there seems to be no satisfactory way to break the circle. A conclusion arrived at some time ago by Toffoli offers a quite general feature of the relation between physics and computing:

> Certain problems are not . . . "solvable" or "unsolvable." Rather, they are to some extent self-referential or *circular*, and as such they don't ask to be solved—they have to be understood and *lived*. In my opinion, "physical limits of computation" and "computational models of physics" are two poles that . . . encompass one of the deepest and most vital of these circular issues. (1982, 174)

Of course, this doesn't absolve us from the need to be critical about our assumptions, about what features of either computing or counting we are willing to accept as given. With this in mind, let's return to Landauer's search for a self-consistent formalism.

Assuming virtually dissipation-free computing to be meaningful enmeshes us, then, in a curious kind of self-enhancing logic. This is because to describe such computing—actually merely to *say* it—one has to invoke the limit construction $V \to 0$. This, in turn, invokes an infinite sequence of integers. The integers are defined to be the result of counting. But counting is a certain kind of computational process, a sequence of physical steps. The obvious question: In what sense is it physically coherent for there to be a potential infinity of physical acts executable by a machine *MU*, each of which is associated with a definite amount of dissipation, energy, and action? And on what logical basis—even supposing the idea coherent and not counter to physics can one *assume* the potential realizability of an infinitude of computational acts as part of one's theory of the nature of any computational act? What epistemological advance is it, in other words, to arrive at a conclusion about the physical nature and limits of the particular informational process we call counting by folding into the very language of one's understanding the notion that numbers go on endlessly?

It seems, then, that Landauer's quest for a self-consistent theory of the physics of computation runs up against a certain barrier. The difficulty is not that Landauer is unaware of the danger of assuming in one's mathematics what is denied in one's picture of physical reality. Unlike the run of physicists, he is, as we've seen, extremely sensitive to the issue, to the lack of contact between an idealized mathematics and an uncooperatively real world; going so far as to castigate the mathematicians for saddling us all with the "given an epsilon, there exists an N . . ." type of inference that produces, among many other things, such illusions of unlimited accuracy as the endless decimal expansion of π. Yet, he is obliged to use precisely this sort of mathematical move, the machinery of the limit—seemingly without noticing, or at any rate without critical mention—when he assents, as he does, to the meaningfulness of virtually dissipation-free computing. The problem is that in spite of his sensitivity he misses a discrete version of the danger he sees so well in the case of the continuum. What haunts the information-is-physical program here is not continuity but nonfinitude: a finite world being theorized and described by an infinite sequence of numbers. Only this time, when the physics is *about numbers*, the worry is far more pressing.

And, since the arbitrarily small is the problem, the only solution is to reject the density property of not only the reals but the rationals too. But to deny the applicability of the density of the rationals is to deny that arbitrarily small rational distances, e.g., 2^{-n} for large n, correspond to physical distances, which in turn is to assert that beyond some region the *integers themselves have no purchase on physical reality* insofar as this reality is held to be measurable.

There is a certain irony in all this. The very subdiscipline of computer science—the physics of computation—one would expect to address the issue of counting arbitrarily far begs the question. And the fact that Landauer explicates the field under the banner "Information Is Physical" only sharpens the irony, since it is the very unphysicality—the idealized, transcendental nature—of the infinitary mathematical formalism that is the heart—or perhaps one should say, immortal soul—of the problem. Thus, if we are to take "information is physical" seriously, then we have to apply it to *all* information, which means recognizing that intelligence and symbolic activity, information about information if you will, is physical.

Such metainformation is the mathematical formalism, the whole symbolic apparatus of number itself. How else do we articulate, describe, define, compare, represent, manipulate, and signify information? Plainly, to talk of "information" in a way that makes it arithmetically quantifiable, to apply mathematics to "bits," to theorize the connection between informational and thermodynamic degrees of freedom in the first place, to define "degrees of freedom," to connect bits to an already mathematized entropy, to articulate a computational account of physics, and conversely to develop a physics of computation, is already to work within a thoroughly mathematized horizon to the idea of what can pass for information. In brief, mathematics functions as the source of all the information about information that the theory of computing seems prepared to countenance as scientific.

To summarize: the proper pursuit of Landauer's goal of a self-consistent account of the ultimate physical limits of computation would demand—as he does but with insufficient rigor—that mathematical formalism itself be sufficiently de-idealized and brought into the field of what we hold to be physical, material, realizable.

This entails inveighing not merely against continuum mathematics but against the uncritical acceptance of endless counting that prevails in contemporary mathematics and physics. And it would prompt one to convert its analysis of infinitism into a view of the whole numbers radically divergent from the current one. In terms of counting, one needs an interpretation of the ideogram ". . ." in the expression "1, 2, 3, . . ." that relates to an instruction to go on counting for as long as it makes sense in a universe like ours. I give an interpretation along these lines in Chapter 5.

POSTSCRIPT: POSTCALCULUS

After centuries of uninterrupted development and progress, the wondrous movable type invented by Johannes Gutenberg has almost disappeared—an industry and way of life obsolesced in a single generation by digital printing. Recently, I tried to imagine Gutenberg's reaction. The occasion was a prolonged encounter with another meltdown, another time-honored practice and its worker denizens becoming out of date and appearing quaint when seen through the lidless eye of the computer screen. This time it was the passing into history of a cognitive rather than a material technology, an illustrious symbolic apparatus, index of the power, prestige, and utility of mathematical thought for the last three hundred years. I refer to the becoming obsolescent and passing away of calculus. Or, at least, calculus as we know it.

In view of the objections to the real numbers and continuum mathematics from physics I have been charting, some such development is to be expected. But it is one thing to engage in a speculative, theoretical exercise, elaborating routes to the possible dethronement of calculus as part of a radical abandonment of the classical continuum and making predictions aimed safely at the near future. It is quite another to come face to face with one's prognostications. My recent experience was anything but theoretical, and the location of calculus's mortal illness has, it turns out, ceased to be in the future.

The source of my enlightenment was teaching mathematics for a year at a private, all-boys, Christian high school; an expensive college-prep outfit, in East Memphis, Tennessee. My job was to induct some of the five hundred boys into the rigors of upper-school mathematics: Algebra II, Pre-Calculus, and Calculus AP (Advanced Placement). Their education included compulsory Latin I (separating the chaff from the eventual, national-merit-winning wheat), a lot of seriously rah-rah sport and coachy jock-slapping, bone-numbing talks in morning assembly (on the pity of poverty, the power of prayer, and the unthinkableness of either atheism or polytheism), compulsory Bible studies (loosely interpreted to include the death of God and the thrills and horrors of moral relativism), and a gentleman's honors system which left lockers unlocked. Seniors—at least those who wanted to be doctors, scientists, or engineers, or who just wanted it on their transcript to get into top colleges—took calculus. Some of them took it, in more ways than one, from me.

Calculus. For an entire school year, I lived the irony of teaching the great blind numbering machine of Leibniz-Newton. A cruel jest to play on someone who'd spent the last half dozen years, exposing, deconstructing, and challenging the theology—Platonic and otherwise—of the mathematical infinite that the bulk of contemporary mathematics (and certainly calculus) rides on. Calculus calculates the smooth curved differentiable deformations of the *continuum*. And the continuum, the straight line coordinate axis of real numbers, as Cauchy, Dedekind, and finally Cantor showed in the nineteenth century, is a deeply infinitistic construct (mysteriously so: even its size in comparison to other infinities is undecidable). So, if gods lurk deep inside mathematical infinity, as I claim, then calculus is presumably a divine science. And here am I, a confirmed unbeliever, heretic, and disser of Plato's heaven, getting a monthly check for teaching the holy ritual of the Church of Calculus to a select few of the nation's young.

But paychecks soak up any amount of irony. I would turn up each day to service the machine: teach; make lesson plans; figure next week's homework; make quizzes, assignments, tests; grade same; work problems; go to assembly; gather gossip and detail on the history and ideo-theological leanings of the institution employing me; proctor study periods; exchange abuse, jokes, and complaints with colleagues; drink a lot of coffee; and wonder—wonder continually—at the point of it all. The point of almost the entire curriculum and school process, but especially the point of teaching calculus, precalculus and pre-precalculus.

I performed sixteen hours a week, dancing in front of the blackboard, waving chalky hands, underscoring, copying, circling and erasing symbols and diagrams (medieval *materia symbolica*). There was a lot of "Hey Doc, why do we have to do algebra? What's the deal on it? We're never gonna use it," and similar beefing, which on the whole I ignored. Occasionally, too pooped and dry-throated to chalk on the board, I'd risk temporary vision defect and write out solutions to problems on the glaring surface of the overhead (nineteenth-century *materia symbolica*). And so I went on cajoling, explaining, torturing, and sometimes illuminating their sixteen- to eighteen-year-old, testosterone-filled minds into abstract disciplined symbolic thought on the way to calculus. The weeks rushed by.

I'd never realized just how manual and paper-intensive calculus is: constantly writing, notating, scribbling, scratching, drawing, and erasing numbers and functions, copying and sketching graphs and diagrams, pushing sequences of expressions and figures about. Of course, such isn't peculiar to calculus—mathematical ideogramming and diagramming go back to *homo mathematicus* and the invention of writing—just a lot more intense. There *homo* was, marking the smooth mud and flattened sand. But these remembered nothing. He then scribbled on rocks, walls of caves, bones, shells, animal skins, himself, others, the sides of stone blocks, monuments, clay envelopes and tablets, bark, papyrus, linen, cotton, and finally on the culminating memory surface of our time: endless pages of refined white paper. That was then. Now, it feels as if all these different writing media, each one improving and über-holing the previous, expanding the envelope and increasing the size of the letter inside it, are being telescoped into a single technological moment, the era of *pre-screen* writing.

A single ten-thousand-year moment? Aren't I overreacting to the absurd necessity of living in the present? No doubt. But alternative vocabularies don't seem to impinge. And surely it's not just me feeling jittery on the edge of the whirlpool: Aren't we all buffeted about and nervously polarized into enthusiastic technophiles or technophobic apocalypts when it comes to the passage from paper to the computer screen? And isn't the nearest thing to most of us, the *body* (the Mobius body, as Rich Doyle twistily calls it) being reinvented at the keyboard, swallowing itself, to come out reimagined and reincarnated, on the virtual, other side of this very screen?

A relative beginning eight centuries ago: the West takes the infidel zero symbol used by the Arabs seriously. Four centuries later, it reinvents the Arabic science of algebra. Immediately, Descartes renders into algebra the ancient, featureless flat plane of Euclid, whose geometry required ingenious proofs that Descartes felt, "exercise the understanding only on condition of greatly fatiguing the imagination." So, Descartes mechanized the imagination by imprinting on this ideal sheet of paper a number grid organized in relation to an ideal zero-zero origin point and two ideal perpendicular coordinate axes. The consequences for science and mathematics were stupendous: points and shapes could be numbered, algebraized; arithmetic could be pictured, diagrammed, and graphed. Two previously separate and independent languages—geometry and algebra—suddenly become intertranslatable. A mathematical Rosetta stone. Out of it came modern analytical mathematics. But finding the pictures of formulas, translating from algebra to geometry, is laborious and requires great skill. In fact, only a small number of familiar functions lend themselves to being actually graphed. Most of the time mathematicians have to work blind. So they invented a way of figuring out the behavior of all those curves conjured up by Descartes's algebra *in advance* of trying to construct, draw, measure, plot, and exhibit them. The apparatus they created for *calculating* and *predicting* increases, decreases, rates of change, optimum values, and critical changes in curvature is the differential and integral calculus. And what an amazingly successful and impressive machine it has been.

Then came the graphing calculator, the TI-82 and such, and the whole gigantic number machine of calculus begins to look like a mechanical artifact from a previous techno-era. I'd been away from math teaching eleven years in body and longer in spirit, and calculus old-style, what I was taught, what the math teachers around me were taught, what ran the revolution of the hard sciences for more than two centuries, what everybody understands by the word, is turning into an appropriately difficult nonliberal art, a weird form of mathematical Latin.

One of my calculus students, Braden, looms into my office. I, trying to speed up my digitonics, am playing with the graphing calculator on my Macintosh. I've just typed in an arbitrary trigonometric expression, hit the GRAPH key and produced a beautiful and totally unfamiliar moiré pattern of loops and apertures. I zoom in and out trying to get the feel of it, aware that mathematics, contrary to its self-promotion as a purely theoretical science of necessary truth, is now—courtesy of the digital computer—also an experimental practice. Of course, the theory/experiment binary is itself problematic. . . . But I don't want to digress. Braden eyes the graph, whispers "cool," then says "get this" as he taps out a function I'd assigned his class, hits DIFFERENTIATE on the menu, and there immediately on the screen is the derivative of the function. His homework—ten minutes of symbolic manipulation, easily screwed up, finicky, requiring total concentration on the details, and needing several years of precalculus and pre-precalculus training—replaced by a foolproof, easy to use, INPUT/OUTPUT box. He grinned and swore he didn't do his assignments like that. Sensible: he knows the calculus testers on the College Board won't let him take his Power Mac into the examination. At least, *not yet*. So we all go through the motions ignoring the fact that classical symbolic practice on paper is being usurped by the instant, granulized alternative, of the pixelated *screen*. From the beginning of writing, *homo mathematicus* wrote by way of curves, shapes, diagrams, figures, indicators, arrays, constellations, ideograms, marks, maps, and symbols. The sundry visual patterns of an intricate repertoire of fine hand-eye motor coordination schema that evolution, in all its unplanned contingency, came up with for tree-dwelling, socially active, opposed-thumbs, gesturing primates. And now the dumb, know-nothing agency behind the TI-82 screen can replace all that finger and thumb dexterity by an instantly plotted graph.

To get a sense of all this I checked out the College Board's position. The board members are answerable to the thousands of high school Calculus AP teachers across the country and, unlike universities, have to take a definite, public stand. The history, as told in a ninety-minute teleconference video aired in 1994 to assuage teacher fears, was revealing. In 1983, before graphing calculators were around, the board ran a limited two-year experiment allowing *scientific* calculators into the AP exam to aid computation. Then they did nothing for several years, hoping perhaps that the calculator phenomenon would go away and leave them to administer calculus-as-usual. By 1989 graphers were starting to be readily available, the 1983 experiment was irrelevant, and business-as-usual a lame hope. They decided to reallow scientific calculators, introduced "calculator active" questions to be answered by them, and resolved to meet what they now saw to be the "real revolution" posed by the presence of graphing calculators. The result: the 1995 Calculus AP exam was the first of a new era. The major part of the exam is still old-time: no calculators, only the familiar algebraic toolkit allowed; the rest, within what they call a GCA ("graphing calculator active") environment, tests previously untestable material designed to incorporate the grapher's ability to answer it. (A side effect: since graphers produce their pictures without any method, students are being asked to *explain* the graphs, thus requiring them and their teachers to become familiar with whole English sentences.) The examiners, as is evident from the College Board's pronouncements, expect the curriculum changes to be only the beginning. Already a new generation of textbooks, reflecting calculator-induced curriculum and methodology changes, is making the rounds. Where will it end?

One can ask the university math departments. Since the flow down from them has established the present curriculum, they should know what "calculus" is to mean. Well, they do, only it's not so simple: a reverse vector has entered the flow. Recall the evolutionary mechanism of *neoteny*, whereby juvenile characteristics are retained in the adults of a species (making human adults like young apes). Here it is a case of cognitive neoteny, a bottom-up vector from the electronically savvy young. The result is a high school driven feedback loop: (1) incoming calc-kids wielding graphers from birth accelerate calculus curriculum changes; (2) College Board Calculus AP panel implements these changes; (3) high school calculus teachers take their cue and focus on the changes; (4) students' environment becomes correspondingly more calculator active; (5) enhanced calc-kids go to college and start another cycle. We're all familiar now with the dramatic and unpredictable changes in quite simple dynamical systems when their output is repeatedly fed back as input. In this case, how many iterations before the whole process stabilizes into "post-calculus"? Maybe when the calc-kids become junior professors. After that, calculus will become "calc," an entirely different—software-led, computer-graphic-driven—game. And after that, it would have to go off the flat screen altogether, stop being confined to the visual and become something kinematic, ambulatory, and tactile, become, in other words, virtual. Mathematics then would be both a prototype and an ongoing product of virtual reality, an idea I touched upon in Chapter 2.

But meanwhile, it's calculus as usual:

Guidelines for sketching the graph of $y = f(x)$:

1. Domain of f. Find the domain of f—that is, all real numbers x such that $f(x)$ is defined.
2. Continuity of f. Determine whether f is continuous on its domain, and if not, find and classify the discontinuities.
3. x-and-y-intercepts. The x-intecepts are the solutions of the equation $f(x) = 0$; the y-intecept is the function value $f(0)$ if it exists.
4. Symmetry. If f is an even function, then the graph is symmetric with respect to the y-axis. If f is an odd function, then the graph is symmetric with respect to the origin.
5. Critical numbers and local extrema. Find $f'(x)$ and determine the critical numbers—that is, the values of x such that $f'(x) = 0$ or $f'(x)$ does not exist. Use the first test to help find local extrema. Employ the sign of $f'(x)$ to find intervals on which f is increasing $[f'(x) > 0]$ or is decreasing $[f'(x) < 0]$. Determine whether there are corners or cusps on the graph.
6. Concavity and points of inflection. Find $f''(x)$ and use the second derivative test whenever appropriate. If $f''(x) > 0$ on an open interval I, then the graph is concave upward. If $f''(x) < 0$, then the graph is concave downward. If f is continuous at c and

if $f''(x)$ changes sign at c, then point $[c, f(c)]$ is a point of inflection.

7. Asymptotes. Horizontal: If $\lim_{x \to \infty} f(x) = L$ or $\lim_{x \to -\infty} f(x) = L$, then the line $y = L$ is a horizontal asymptote. Vertical: if $\lim_{x \to a+} f(x)$ or $\lim_{x \to a-} f(x)$ is either ∞ or $-\infty$, then the line $x = a$ is a vertical asymptote.

Seven steps, three and a half centuries of accumulated techniques for curve sketching, developed laboriously and carefully and with increasing rigor and complexity, collapses into a why-bother-with-it-all brandishing of a TI-82. A few one-finger keyboard beeps and the graph of your function draws itself on the small, few-thousand-pixel screen in your hand. Layers of carefully refined cognitive know-how behind those guidelines sent into oblivion by an $80 hardware/software computing device, which is only the first and most primitive of its kind. I told my students they'd entered an interregnum: ten years ago no graphers—classical calculus and pre-calculus from Descartes was king; ten years from now the grapher will rule, determining the curriculum and the scope of the subject. But now, between kings, they had buckle down and learn soon-to-be-obsolete skills.

Many vectors lie inside this shift from classical calculus to calc. There is the transformation from a discipline dedicated to accumulating theoretical knowledge about ideal curves, to one using modeling software to investigate real curves; the shift from primate hand-drawing with all its filiations to human gestures to mechanical plotting of pixels; the move from the point on paper, ideal classical object, originally defined by absence now as an infinite limit, to the materially present pixel on the screen; the move from the continuous Euclidean-Cartesian plane, infinitely extendible, uniform, and featureless to the discrete and bounded grid of cellular elements; there is the transition from printing signs that have originated outside the medium of inscription to exuding signs manufactured as an integral part of their medium.

Of course, relative to the blizzard of cultural change we are moving through, the reconfiguration of calculus into calc is little more than a snowflake. Even within mathematics, the vectors I have listed are a metonym of a wider transformation. Calculus's fate, as we have seen, is bound up with the classical real continuum; and this in turn is bound up with the fate of the concept of infinity as this confronts the deeply finite materiality intrinsic to the digital computer. This is not to say, however, that the changes represented by these vectors result in any simple substitution, an overthrow of the old by the new. Rather, they point to the creation of a de-theologized alternative to the classically endless finite, a way of coexisting with the classical conception that refuses to be categorized as inferior to it. The upshot is that "finite" becomes not that which falls short of a prior—already and always there from the Godhead—infinite, but that which includes us, its cognizers, from the biological and material outset. Correspondingly, as I indicate in Chapter 5, "infinite" becomes reinterpretable precisely in terms of this materiality.

CHAPTER 4

God Tricks; or, Numbers from the Bottom Up?

Let me start with a fine book, *The Origins of the Modern Mind* by Merlin Donald (1991). In the concluding section, called "Exuberant Materialism," Donald places his account of the evolution of culture and cognition within a burgeoning and newly self-confident materialism, part of what he calls a "neuroscientific apocalypse." The aim is to construct a biological morphology of intelligence able to lift itself free from the "tortured logic and compromises" that have characterized discussion of the mind/brain relation in the twentieth century, and to reach toward a scientific understanding of human consciousness—an understanding, Donald emphasizes, going beyond the familiar Darwinian naturalism to depict humans as "symbol-using, networked creatures, unlike any that went before us." "We act," as he puts it, "in cognitive collectivities, in symbiosis with external memory systems. As we develop new external symbolic configurations and modalities, we reconfigure our own mental architecture in nontrivial ways" (Donald 1991, 382). The claim is striking. We are, it seems, constantly rewiring our brains as we go along, physically manufacturing our mentality via the symbols, cognitive tools, and technosemiotic apparatuses that are the cultural-mental, ideational-achievements of these very brains.

For some years I have been concerned with one particular external memory system or symbolic configuration—mathematics: mathematics understood as a technosemiotic apparatus, that is, as a *discourse* with ideograms and diagrams, as a *technology* (tool for reasoning, thinking, predicting, imagining), and as an *apparatus* (system of writing and manipulating material symbols). In addressing these issues one quickly encounters the mind/brain relation and its forebears and cognates—mind/body, psyche/soma, mental/material, idea/thing, heaven/earth, spirit/flesh, signified/signifier, and so forth. The relation is of course one of the founding binaries of Western culture; no surprise, then, that it impinges on and problematizes what we mean by "doing mathematics" and by "counting," and that this occurs as soon as we write down the familiar progression of ideograms "1, 2, 3, . . ." symbolizing the so-called natural numbers and start to ask semiotic questions about it. The primary difficulty concerns not so much the character and status of the individual numbers themselves, as the interpretation of the continuation symbol ". . ." operative on all of them: they form a progression. How are we to give a linguistic account of unlimited progress of numbers? What would be the appropriate semiotic response to infinity? How can we witness it in terms of written signs? And, perhaps more opaque at this stage, what has the mind/body relation to do with the problem? I shall return to these questions later.

A second difficulty can be located in the inscription of the mind/body binary within classical Greek formulations of number. One of these has the mind giving rise to what was called *arithmetica* (numbers in heaven, ideal objects of philosophical contemplation); and one has the body giving rise to *logistica* (numbers on earth, lowly things to calculate with). Derivative and supportive of this is the whole Platonic philosophy of mathematics, which contends that mathematical objects—points, lines, numbers, spaces, functions, surfaces—exist in some ideal realm, independently of any human presence, activity, or knowledge; prevalent among mathematicians, scientists, and the general populace, it is a belief strongly and confidently upheld.[1]

For Platonists, then, mathematics discovers truths about this pregiven external realm in just the same way that (realist-conceived) science studies the world of so-called external reality. And for both, language is no more than a transparent, inert medium whose features, interesting though they might be, have only a surface, marginal, and in principle always eliminable bearing on what is "out there." In short, for Platonic realism, language is ultimately nothing other than naming.

According to this, as we saw in Chapter 1, numbers are objects, signifieds, which preexist the names—numerals, signifiers—that we assign to them. What would it mean to challenge the hierarchy and the separation of terms of this hallowed number/numeral coupling? Is it not the case that even anti-Platonists (such as mathematical constructivists) are not prepared to give up on an autonomously generated succession of integers outside and before language? How could the picture of a preexisting world of numbers and a posterior assignment of names be overturned? Could numbers somehow owe their existence to that which merely notates them? Is this not absurd: Do not the signifiers "5," "V," "five," "funf," "cinque" name the *same* number—one that, however it is referred to, exists prior to and independently of the attachment of such signifiers to it? Does not any alternative ask us to believe in not one, pregiven progression of natural numbers, but many artificial progressions, each inseparable from the particular system of signifiers used to bring it into existence? For almost all mathematicians this—and, by implication, *any* attempt to dissolve the separateness and inviolable priority of the signified over the signifier—would go beyond sense.

But why should the situation of mathematics be so special and singular—so intransigently different from *language*—in relation to this possibility? Is there nothing to be gained from two decades of poststructuralist, deconstructive theory devoted to precisely such a dissolution? Thus, consider a familiar linguistic rather than mathematical example: the opposition of literal and figurative. Here again one has a privileged term, a presumed priority—the literal serving as the ground from which figurative language, such as metaphor, is traditionally derived. But, as Jonathan Culler observes, such an "asymmetry turns out to be unstable, and as one explores the logic of the situation further, one discovers that the term treated as secondary and derivative can be seen as basic" (1981, 206). The result being that the very ability to proclaim an absolute separation becomes untenable, and one is left with a reconstituted reading of the literal/figurative opposition as a codependent or codetermined coupling that denies the possibility of an atemporal structure of opposition. The move is quite general, and Derrida, who calls this decentering and relocation of a supposed supplementary term the "logic of the supplement," has famously enacted this logic on the speech/writing opposition and, more to the present point, on the signifier/signified binary that constitutes Saussure's understanding of the sign.

Trying to use Derridean deconstruction, however, with its specialized lexicon of "logocentrism," "supplement," "trace," "*différance*," "presence," and so on, *directly* in order to critique the mathematical binary of number/numeral is not a good idea. One has to go indirectly and from behind as it were—deconstructing (although that is not the word I would prefer to use) mathematics within its own already highly specialized terms.

Let me leave that hanging and put mathematics in abeyance for the present, and go instead on a curve through certain projects in robotics, evolutionary and developmental biology, artificial intelligence, cognitive science, and social anthropology. Thirty-five years ago Warren McCulloch wrote his celebrated *Embodiments of Mind*. Since then, especially in the last decade, a slew of books echoing, advancing, and reconfiguring his assault on the mind/body binary have appeared, their titles and subtitles—*The Embodied Mind*; *On the Matter of the Mind*; *The Body in the Mind*; *The Biology of Mind*; *The Mindful Brain*—looking to exhaust the list of possible anti-Cartesian slogans.[2] The projects that I am going to talk about are part of these efforts and are linked to one another by their allegiance to a set of key terms, principally, "situated," "enacted," "embodied," and "codetermined," and their characterization in terms of a certain move—at once a metaphor, an algorithm, a principle—described as bottom-up (body) as against top-down (mind).

Before I start, a note of caution on the "body." Recently, Katherine Hayles, in an essay on materiality, suggested that "one belief from the present likely to stupefy future generations is the postmodern orthodoxy that the body is primarily, if not entirely, a linguistic and discursive construction" (1993, 147). I suspect future historians will be spoiled for choice when it comes to being stupefied by our present. Nevertheless, she is surely right to emphasize the servitude to theory and high-mindedness that excludes the biologically material, the experiential, the corporeal, and the enactive from what we mean by embodiment. No such exclusions are intended here.

Let us start from a certain parallel between forms of immersion and separation: minds inside and distinct from bodies (organisms), and bodies immersed in and distinct from the world (environment). Pursuing this, one approach to the binary of mind/body would be through the biological opposition of organism/environment. The prevailing evolutionary picture is the familiar neo-Darwinian synthesis, which posits a genetically determined organism immersed in and selected for by a preexisting and effectively independent environment. Recently, several proposals for rethinking this relationship have been suggested. Perhaps the least contentious is that which foregrounds the phenomenon of *coevolution*. Here, species are quite literally parts of each other's environments—hence the ecological reciprocity of prey and predator couplings, or the coevolution of insects and flower-bearing plants, and so on. And this applies not only to species: DNA and RNA themselves are now thought to have come into being and evolved reciprocally in relation to each other.

Others urge a more theoretically radical, full-blown codetermination between life-forms and the environment in, through, by, on, which (the prepositions here are part of the point) evolution takes place. As Richard Lewontin puts it:

> The organism and the environment are not actually separated. The environment is not a structure imposed on living beings from the outside but is in fact a creation of those beings. The environment is not an autonomous process but a reflection of the biology of the species. just as there is no organism without an environment, so there is no environment without an organism. (Quoted in Varela, Thompson, and Rosch 1991, 198)

By this is meant that in a sense our bodies enfold past environments and vice versa. And this not only within species at the level of populations, but also in the development of individual organisms; so that, far from being separate and independent, the organism, in its passage from genes to adult, and the environment that mediates this passage are codetermined and mutually enfolded. In Susan Oyama's words, "Genetic and environmental influences are made to be interdependent as genes and gene products are environments to each other, as extraorganismal environment is made internal by psychological or biochemical assimilation, as internal state is externalized through products and behavior that select and organize the surrounding world" (quoted in Varela, Thompson, and Rosch 1991, 199).

This refusal of a preformed and independent world is given a cognitive dimension by Francisco Varela, Evan Thompson, and Eleanor Rosch in *The Embodied Mind*. For them the crux is that "cognition is not representation but embodied action and the world we cognize is not pre-given but enacted through our history of structural coupling" (Varela, Thompson, and Rosch 1991, 200). By "representation" they mean mirror-realism, and rejecting it forms part of repudiating the core claims of *cognitivism*—the prevailing mind-as-a-computer paradigm in artificial intelligence. Against the model of thinking as rule-based information processing and manipulation of symbols mirroring an autonomous, preexisting reality, thinking becomes more like perception-in-action, where what we do guides thought rather than the reverse. And instead of a functionalism that has the thinking device (hardware) as irrelevant to what it thinks (software)—leading to such eye-widening wonders as downloaded minds and backup copies of the self—one has the organic and messy wetware of neuronal pathways. In short, action and intelligence are to be seen not as the *result* of rational thought—such as representation, beliefs, symbols, and planning—but as the cause and *source* of rationality itself. This means that cognition, in its moment-to-moment operations as well as its overall structure, is not controlled by a plan from the mind above, but rather bubbles up, as it were, hesitantly and blindly from the body below.

For Gerald Edelman in *Neural Darwinism* (1987) and Jean-Paul Changeux in *Neuronal Man* (1986) such an upward vector, working via selection, operates throughout the evolutionary morphology of the brain. Edelman cites the immune system as the motivating example. In order to make antibodies that match foreign proteins, the immune system can operate in two opposed ways: either it can, in advance and perhaps randomly, produce millions of different antibodies and then increase production of the one that matches the alien protein, or it can wait and see, checking out the structure of a foreign protein as it arrives and using the information it extracts to manufacture the appropriate antibody. It turns out that the second strategy, rational and surely more intelligent, is not the one used; the immune system, we learn, uses the principle of selection applied to the mindless proliferation of possibilities. The thesis Edelman promotes is that a parallel phenomenon (only vastly more complex) involving not antibodies but groups of neurons takes place within the organism at the level of perception, memory, cognition, and eventually consciousness. In other words, selection from below, rather than rational guidance and information processing from above, is the mechanism by which the mind/brain system—as well as the immune system itself—came into existence and evolved.

The reliance on selection here is a crucial move. Selection, as Edelman points out, marks the methodological gulf between physics, whose particles are identical, and biology, where individual variation within populations, far from being an unwelcome fuzziness to be idealized away, is the sine qua non of all evolutionary change. Thus, not only is computer-inspired functionalism as a matter of fact explanatorily inadequate, but the computer, operating like physics on identities as it does, is in principle misguided as a model of mind. "Selection," Edelman says, "contrasts starkly with platonic essentialism, which requires a typology created from the *top down*; instead, population thinking states that evolution produces classes of living forms from the *bottom up*" (1992, 73; emphasis added).

It is precisely this bottom-up versus top-down figure, conceived as an overriding methodological principle, that MIT robotics engineer Rodney Brooks nails explicity to his mast. But before I introduce his work, let me enlarge on the figure itself. The range of its surface occurrences—from the Pentagon's recent expenditure review, to attacks on the autonomy of school boards, to corporate downsizing, to current work in artificial intelligence and neurobiology—suggests a deep-lying and apparently much needed image. What is the preference or prejudice at work here? What would top-down over bottom-up (to invert the order)—as intellectual method, organizing metaphor, or cognitive style—*mean*? One might gloss it as the ranking of the global, panoptic, abstractly analytic over the concrete, limited, and locally synthetic; of posterior description, morphology, and structure over history, evolution, and genesis; of plans over objectives and goals; of general laws over incidents and cases; of context-free reason over situated knowledge; of realist truth over constructivist emergence. But the figure is also reflexive and applies at once to descriptions of itself: the summary I have just given (which is perhaps less than helpful as an explication) is very much a top-down take on the bottom-up / top-down difference. So let us try it the other way—through a bodily situated example.

Suppose you drive in both France (or the United States) and Britain (or Japan) and wish, for whatever reason, to avoid getting killed, particularly through a head-on collision. Then you need to know a simple top-down rule: drive on the right in France and on the left in Britain. But what if you cannot remember which way around the rule goes; or (common in academia) you get your left and right mixed up; or, jet-lagged in the middle of a dark nowhere, you get confused as to whether you are in Britain or not—how are you going to drive and stay alive? One possible response: get off the road; if you are unable to remember a one-line rule, do not know your right hand, and forget which country you are in, you should not be driving. Well, don't despair: here is a bottom-up procedure you can follow. When entering a road you look through your own—the driver's—window to see oncoming traffic, and you join it; on the road, you make sure you are always sitting in the middle (equivalently: the roadside isn't rushing past *next* to your window). That is all you will ever need. Observe: the suggested strategy is about as mindless as you can get—you do not have to remember an abstract rule, only a procedure tied to your physical situation; a knowledge of right and left is not needed; and the strategy works for whatever country you are in even if *you* don't know which. Of course, there is a joker: you need to be in a French car in France and a British one in Britain; the position of the steering wheel is obviously crucial.

Now that I have helped you drive abroad, let me return to our theme. How is it, Rodney Brooks asks, that after thirty years of prodigious effort, orthodox or classical AI—the current paradigm—has produced robots that can do no more than make their way clumsily across a carefully prepared, never varying toy world, even though they incorporate any amount of fancy hardware and have access to vast computing power running at state-of-the-art megahertz speeds—whereas a bee, for example, employing truly ancient technology, with a tiny amount of on-board computational power running at a few kilohertz, can navigate a previously unseen, noisy, and possibly hostile environment as well as perform a host of tasks, from foraging and communicative dancing to returning to the hive?

Has evolution, then, a better purchase on robotics than AI orthodoxy? In the past half dozen years Brooks has been demonstrating how and why the answer might be a very sure *yes*. In a paper punningly entitled "Intelligence Without Reason" he dismisses traditional AI for "trying," as he says, "to tackle the problem of building artificially intelligent systems from the *top down*" by relying on the introspective notions of thought and reason and their computerized interpretations via information processing, conscious planning, and problem solving; instead, he advocates an approach based partly on biology, on an older cybernetic tradition, and on engineering from first principles—one that starts "from the bottom up concentrating on physical systems . . . situated in the world, autonomously carrying out tasks of various sorts" (1991a, 1). This has fundamental design implications: instead of a centrally controlled hierarchy of connected systems—a robot with a "mind"—Brooks uses what he calls "subsumption" architecture: robots are built in layers consisting of simple machines that are minimally interconnected, each layer independent of any above it and each, by autonomously controlling its own pattern of sensors and actuators, engaged directly with the world.

The approach has paid off: using a half dozen or so layers Brooks has constructed a variety of robots, from six-legged protoinsectoids that scuttle successfully across littered desktops to machines that roam his laboratory looking for and collecting empty soda cans. (Aside: one should be wary of locutions like "looking for": each internal state of Brooks's robots lasts no longer than a fraction of a second and any intentionality is therefore, as he is quick to point out, in the mind of the beholder.) His immediate goal is to reach the level of full insect intelligence, which he estimates would require some fourteen layers. Given the coevolution with trees, flowers, and insects of our evolutionary ancestors, building a bee (or bee-oid, one should say) would surely be an achievement, with implications well outside pure robotics.

Brooks enunciates "embodiment" and "situatedness" as the organizing principles of his bottom-up, behavior / action–based approach to building intelligence. In his usage, the terms are separate. *Embodiment* means that "the world grounds regress" (1991a, 16), which entails that what his robots do is inseparable from the details and constraints of their construction, thus ensuring that their own physicality, rather than prior desiderata, will govern whatever symbolic meaning is to be attributed to their activity. One can see embodiment as a thinned-out version of individual human agency: a contained, palpable projection of body behavior onto the space of an "it" with a small number of degrees of freedom; for example, a paint-spraying robot on a production line.[3] *Situatedness*, on the other hand, embraces for Brooks an ongoing relation to a changing world that has surprises and a history: instead of a pregiven plan relating to a representation of an oversimplified environment, a situated machine or agent must use its immersion and changing perceptions in the here and now as the basis of its response; "The world is its own best model" is how Brooks expresses it (1991a, 15). Recall our driving example where incorporating the world—oncoming traffic, steering wheel—was crucial. More generally, Brooks gives an airline booking system as an exemplar, making situatedness a thinned-out version of social agency—a dispersed, invisible ticket-processing "they."

Brooks's attachment to the bottom-up procedure is also performative, ruling the description as well as the content of his approach. Thus, not only mind—problem solving, central control, representation—is subordinated within his model of intelligence but also its sociocultural correlates—philosophy, abstract thought, theory—are likewise invoked by him on a minimal, need-to-know basis. Such double mindlessness is a powerful head clearer, enabling the binaries of real/artificial or natural/machine intelligence to be productively opened up.

One aspect of these binaries—"the problem of human-machine communication," as she puts it—provides the subtitle of social anthropologist Lucy Suchman's book *Plans and Situated Actions*. In order to think our connection to machines, Suchman argues the need to set up a theoretical framework for understanding how action always is what she calls "situated" and, correlatively, how we understand and use *plans*. It should be evident by now that the theories of cognition and robotics just discussed reject plans—detailed representations, predetermined courses, or blueprints of action—as the principal or even necessary component of what we call intelligent behavior. Suchman's contribution—and her concern is neither robotics nor cognition, but rather socially constructed meanings—is not that plans are absent from purposeful social activity, but instead that our engagement with them vis-à-vis communication and shared understanding is other and more complicated than common sense (and hence cognitivism and orthodox AI) imagines it to be. In her words:

> The confusion in the planning literature over the status of plans mirrors the fact that in our everyday action descriptions we do not normally distinguish between accounts of action provided before and after the fact, and action's actual course. As common-sense constructs, plans are a constituent of practical action, but they are constituent as an artifact of our *reasoning about* action, not as the generative *mechanism of* action. Our imagined projections and our retrospective reconstructions are the principal means by which we catch hold of situated action and reason about it, while situated action itself, in contrast, is essentially transparent to us as actors. (1987, 38–39)

Such a reversal and displacement—plans as action-driven rather than actions as plan-determined—requires a radical repudiation of classical (Durkheimian) and therefore much current sociological orthodoxy—a repudiation that, as Suchman observes, was prefigured by G. H. Mead seventy years ago and extended some thirty years later into the basis of what is now termed "ethnomethodology." According to classical sociology, there is "an objective reality of social facts" that individual actors obey or conform to, received norms that determine our actions and attitudes. Sociology is the would-be science of this objective realm. To reverse this is to hold that what is called objective social reality is the *problem* of sociology—how did and do we manufacture objectivity?—not a given from which the subject starts. Suchman sees her notion of situated action as deconstructing a classically sanctioned binary: producing a third way between a behaviorism that reduces the significance of actions to uninterpreted body movements and a mentalism that ranks actions within the physico-social world as secondary and epiphenomenal.

Suchman develops this insight in different directions: from a

Heideggerean recognition that action becomes visible—and hence subject to rational scrutiny and after-the-fact narrativization—only when dysfunctional or otherwise interrupted, to an insistence on the indexicality of signs, the fact that "the sign is actually a constituent of the referent" in any account of mutual intelligibility—particularly at the interface between persons and machines (1987, 62). It is the latter that enters into the practical application of Suchman's understanding of the idea and mechanics of situated action. In fact, Suchman is an employee of the Xerox corporation at Xerox PARC, so let me elaborate the application by moving to John Seely Brown, director of PARC, and his prize-winning piece in the *Harvard Business Review* that puts Xerox's employment of anthropologists like Suchman in a certain corporate context.

Brown's paper opens with a startling declaration: "The most important invention that will come out of the corporate research lab in the future will be the corporation itself" (1991, 102).[4] By this he means us to understand that it is new organizational and technological architectures, corporate reconfigurations conducive to product innovation, rather than the products themselves, that need to be invented. "Corporate research must," he insists, "reinvent innovation." Why is this necessary? The short answer is that the innovation in question is being occluded by the structure of the corporation itself. Unquestionably, like the military, the Catholic Church, the modern state, prisons, universities, schools, and hospitals, corporations are top-down outfits with power and control flowing from the CEO (pope, ruler, governor, principal, provost, and so on). Changing this would surely be revolutionary: recalling the Trotskyist slogan of the early 1970s—"Invert the vector: power to the people"—one might well take Brown for a corporate anarchist. But the business community need not fear: it is the vector of ideas, novelty, information, and invention, not that of power, that Brown wants to invert.

How did the need for this inversion become evident? "Some of our most important research," Brown writes, "has been done by anthropologists" (1991, 108). Enter Suchman. Thus, consider the question never far, one supposes, from corporate lips: "What do workers *do*?" Ask the CEO, the management, or the workers themselves, and the answer is more or less in line with the *job description*, the formal, generic characterization of procedures that are to be followed—and indeed, this is what Suchman got when she put the question to office clerks at Xerox. But when she observed them, a very different answer emerged: the clerks were not following the procedures in the manual but instead were using a variety of informal methods to get the work done. They were, in Brown's description, "constantly improvising, inventing new methods to deal with unexpected difficulties and to solve immediate problems" (1991, 108)—being, in other words, innovative in ways nobody, including themselves, had been aware of (or, at least, able to articulate). Like plans in Suchman's analysis, the job-manual description was either after or before the fact, useful for comparison, levels of remuneration, insurance, and the needs of management, but neither a determiner nor an accurate description of working reality.

A similar finding emerged about tech-reps, the people who go out and service Xerox machines. Since the rate at which they can be trained puts an upper bound on the speed of new product introduction, their education became the focus of Suchman's ethnomethodological eye. Here again a crucial disparity between formal and informal, situated and unsituated action, came to light. It was not, apparently, through Xerox's well-funded, state-of-the-art teaching methods, but through casual, unofficial, locally produced *stories*—coffee-break gossip and situated on-the-job narratives of breakdowns and unforeseen glitches—that they learned what they needed to know. Circulated informally among themselves, the self-education of the reps, like the innovations of the clerks, was invisible to management.

It was invisible, that is, to management governed by the top-down vector of the traditional corporation—and, more important for Brown, even when made visible, inaccessible because of the local, unplanned, bottom-up generation of such information. And this is the burden of Brown's song: the corporation's necessary self-reinvention in order to harvest its own, internally produced innovation, a reinvention obliging the corporation to dissolve aspects of the management/worker binary and involve itself in a process of mutual determination that he calls *coproduction*—first coproducing itself with its own workforce, and ultimately coproducing with its customers the very technology and potential applications that it sells to them.

Let me summarize. We have gone from the organism/environment binary reconfigured as a codetermined coupling within post-Darwinism, through the embodied minds, cognition-in-action, and neuronal selection of postcognitivism, to the mindless robotics of a situated and embodied postclassical AI, to, finally, the situated action of a post-Durkheimian social anthropology. In each case the move was explicitly formulated by its adherents as one that refused the flow of ideas, choices, information, decisions, or control from above, refused a theory or perspective laid out in advance of what it would apply to, in favor of what can only be called an *emergence* of action, intelligence, and conscious awareness from below.[5]

But there is an important source of bottom-upness that I have not yet mentioned—one more concerned with cultural-political rewriting than with furthering scientific or corporate instrumentality, and one that has become so omnipresent as to be no longer identifiable within the top/bottom matrix of its original emergence. I mean of course feminism, which in the above list would be glossed as post-patriarchy, would appear as the attempt to completely reconfigure Western culture from the bottom-female-up, as well as being literally and figuratively the mother of all manner of embodiments. It is quite impossible to do more than mention feminism here; but I will make do with a single image taken from one of Donna Haraway's feminist manifestos, relevantly titled "Situated Knowledges" (1991). For Haraway, being situated involves privileging what is always and necessarily a "partial perspective," and her name for the rhetorical sway of the unsituated, the totalized, the disembodied—at least as it appears in science's claims to unlocated, context-free, objective knowledge, spoken in what the philosopher Thomas Nagel calls a "voice from nowhere"—is the "god trick," which—given the direction and altitude of heaven—is surely the most sanctioned, sacred, and deeply defended top-down move in Western culture.

No knowledge is thought to be more context free and universal than mathematics; universal not just for all people, in all cultures, for all time, but also for all forms of intelligent life anywhere in the universe; making mathematics the obvious language to contact alien life-forms. Thus, at the dawn of the space age, Hans Freudenthal converted Leibniz's mathematical project of a *characteristica universalis* into a *linguistica cosmica*—"Lincos" as he called it—in order to communicate with said aliens. First one would teach them how we write numbers, then arithmetical operations, then algebra, then coordinate geometry, calculus, and so on, up to the equations of modern physics and ultimately, given the foundational role of physics, the whole of science and—given its foundational role—all human knowledge. In that way they would know what "stage" we had reached. This idea was then modified to include sending digitized drawings of various fundamental intellectual icons; a proposal not far removed from the turn-of-the-century scheme to signal to Martians by way of a vast diagram of Pythagoras's theorem laid out on the surface of the earth. Listening for and sending messages to aliens—suitably Cartesianized into disembodied ETIs (Extra Terrestrial Intelligences)—by scientists like Carl Sagan and others was a sport of the 1970s and early 1980s. We are still waiting for the alien Other, imagined (insofar as such a project could imagine it/they to be) as disembodied rational thought-processor, to send electromagnetic waves into the giant, federally funded listening dishes, silent radio ears waiting poignantly in a forest clearing inside Puerto Rico.

But might not the alien (and their "mathematics") be more other, more incongruous and unthought in its difference from humans, than this. Or to put it bottom up: humans are embodied, fleshed, and corporeal in very particular, not to say peculiar, ways. Unlike the imagined ETIs, humans are biologically incarnate and messily put together beings who arrived here on planet Earth at the end of a long chain of contingent biological links. According to evolutionist Stephen Jay Gould the contingencies were so ever-present and inescapable that the process could have gone by chance, hazard, biological noise, random happenstance, evolutionary caprice—billions of other ways: we are just one of the many possible outcomes. Pity the poor alien, then, being contacted, who has to be able to handle any one of these outcomes: whispering lists of prime numbers into the giant ears in

Puerto Rico for us, performing heaven knows what other contortions to contact any of the other billion life-forms that might have been us. But I digress.

I want to use Haraway's image to link my curve through bottomuppery back to mathematics. Where in mathematics, one might ask, is the god trick turned? The short answer is at and through infinity. A longer answer would reveal two god tricks—Plato's and Aristotle's—corresponding to two ideas of number and hence of the infinite: Plato's corresponding to the numbers' being and always having been "out there" (actual infinite), and Aristotle's corresponding to an endless coming-into-being of the numbers (potential infinity). The latter is logically weaker and taken as metaphysically unobjectionable; there is more mileage, then, in exposing the god trick it rides on. To do so one needs to say how god is smuggled into mathematics incognito inside the innocent-looking continuation symbol ". . ." that we write after 1, 2, 3.

Let us ignore the usual job description given of mathematics (exercise of pure reason, pursuit of objective truth, free play of the mind, and the like) and operate ethnomethodologically. We observe that mathematicians spend their time scribbling and thinking: writing or manipulating or disseminating (to themselves and others) a prodigious range of symbols, as well as thinking about all manner of imagined worlds and the objects/processes within them. The two activities are obviously related, although how and why seems opaque if not mysterious. To go further one has to look more closely at the kind of writing that takes place, at its rules, protocols, and symbolic conventions and the way these semiotic features hook onto—feed and are fed by—the mathematical imaginary. A closer look reveals two purely grammatical features of rigorous mathematical texts: they are dominated by the imperative mode, amounting at times to little more than networks of injunctions; and they exclude all indexical signs such as "I," "here," "now," and "this." Two questions, then: Who or what utters these injunctions, and to whom are they uttered? And what does it mean to lack the kind of subjectivity and self-location that indexical terms make available? These questions lead inevitably to the interiority of mathematical practice, to the demand for a quasi phenomenology of symbol manipulation that operates from the mathematician up.

The basis for such a phenomenology can be found in the semiotic writings of Charles Sanders Peirce. Elaborating a suggestion of his, one can describe, as we saw in Chapter 1, mathematical reasoning as a species of waking dreams or *thought experiments*. Such "reflective observations," as Peirce called them, are like those we use every day when, in imagination, we manipulate a proxy or surrogate of ourselves in order to figure out what would happen if we—our material selves—did the thing in question. For mathematics, the imagined scenario is a world conjured into being through written signs, and the thing being mentally tested is a mathematical statement, which, on this view, becomes a prediction about the mathematician's future encounters with signs. In other words, mathematics is a rigorous inscriptional fantasy: the insistence on writing determining what can be legitimately imagined, and the ongoing process of imagining controlling what mathematicians can meaningfully and usefully write down.

How, according to this, are we to understand counting? In particular, how do we give meaning to the repeated and always repeatable adjunction of an idealized mark coded into the interpretation of the ideogram ". . ." of endless Aristotelian counting? Who is, could be, or will be making these marks? Evidently not *us*—easily bored, error prone, confused, and mortal—but the surrogate self of a thought experiment. Such a self is idealized to avoid the vicissitudes of actual counting: we require it to perform thought experiments, not real ones. Now, if it is to be persuasively predictive for us, then it must be a human simulacrum, an idealized self that resembles us—we who send it on its journeys—in some essential way. Otherwise, why credit its activities as meaningfully related to, impinging on, and affecting us? But idealized in what way, to what extent? What about us, as counting, signifying agents, is essential? Should we—and the answer determines whether the god trick, Aristotle-style, is turned or not—insist that our embodiment, our incarnation within the physical and material universe, be a feature of any human simulacrum deserving that name?

If we answer no, then we conjure into being a totally disembodied agent, an imagined surrogate of ourselves that operates outside the regimes of time, space, and any kind of material presence—a being unsituated and free of any context we can name, godlike in its ghostly and untouchable otherness. This negative response is the only one that contemporary infinitistic mathematics *can* give if it wants, as it does, the means to count endlessly. This is because if we answer yes, if we insist that any surrogate of ourselves be—in no matter how idealized a way—embodied, then it will be part of the regularities and limitations of the universe we all inhabit. It will operate in accordance with time, space, and materiality in the presence of noise and error and will, therefore, be unable to count endlessly.

One choice, then, confirms contemporary Aristotelian infinitism with its endless progression of numbers. The other choice leads to a nonclassical, non-Aristotelian understanding of iteration, a mode of repetition that I call "counting on non-Euclidean fingers." I shall try to spell out what such a counting from the bottom up produces in the way of a new configuration of "number" in the next chapter. Suffice it to say that the resulting numbers exhibit an organization and a set of mathematical questions radically different from those presented by the conventional picture of number. What can instead be emphasized here is the overall strategy involved, the idea built in from the very start, of the embodiment, the materiality of the counter.

We must think of this materiality in two senses. The term "counter," like "signifier," denotes both the agency—the one-who-counts, the one-who-signifies—and the object used—tally mark, sign vehicle, and the like. To insist, then, on the materiality of the signifier cuts the ground from under both the god tricks in mathematics. No longer are the whole numbers given, natural, and before us, the work of a god-mind whose plan for them, in place from the beginning, preempts all understanding of what it means to count. Numbers no longer simply *are*, either in actuality or in some idealized potentiality: they are materio-symbolic or technosemiotic entities that have to be *made* by materio-symbolic creatures. They and their arithmetic are always part of the larger and open-ended human initiative of constant becoming—an enterprise never free from choice, contingency, the limits of our (always material) resources, and the arbitrariness of history. To understand numbers in this way is to move toward an ecologically more sensitive—greener, more biological—mathematics; one in which neither Aristotle's prime mover nor Plato's divine mind makes its infinite appearance, since in it numbers have to be grasped bottom up from the living body of the counting subject.

CHAPTER 5

Counting on Non-Euclidean Fingers

In several places so far, I've mentioned the possibility of an alternative kind of number with its own arithmetical laws radically different from the usual, classical understanding. I have called these numbers and their arithmetic *non-Euclidean*. What is characteristic about them is their connection to, or rather their emergence from, primary materialities; the materiality of the one-who-counts, the materiality of the counting process, the materiality of writing and signifying; materialities that become the site of palpable production of "numbers" as soon as numbers are understood as made within this universe and not found by some mysterious process in a necessarily inaccessible and always receding mathematical heaven. Following are two attempts—which couldn't be more different in tone, style, intent, and accessibility—to exposit such production. Each takes a different aspect of the material world as its starting point.

The first began life as a self-interview and ended as a much edited, mock-Galilean dialogue—a format that forces it to be direct and to the point with little room for qualification or subtlety of exposition. It manages, as a result, to communicate an image of what non-Euclidean numbers might be (and why they are natural objects to arise in the digital era) without getting caught up in too much theoretical and abstract justification. The second, on the contrary, is abstract and theoretical. It was solicited for a conference on the work of Gilles Deleuze and Félix Guattari and is philosophical in the way that their writings demand. However, one doesn't need to have read them to get the main idea behind the essay, which is to make a connection between non-Euclidean numbers and the minimal swerve or *clinamen* posited by Lucretius as the original deflection from pure, unchanging motion of atoms in the void. Specifically, I offer an account of the *clinamen* that understands it as the uneliminable presence of the materiality of the universe; one which when incorporated at the absolute beginning of the counting process (namely, the passage from zero to one) separates the non-Euclidean world of iteration from the pure, identical, and endless repetition of its classical counterpart.

THE TRUTH ABOUT COUNTING

> "Can you do Addition?" the White Queen asked. "What's one and one and one and one and one and one and one and one and one and one?"
> "I don't know," said Alice. "I lost count."
> "She can't do Addition," the Red Queen interrupted. "Can you do Subtraction?" —LEWIS CARROLL, *Through the Looking-Glass*

The hour was cool, and the sage Kronos and his eager young mathematical protégé, Simplicius, had eaten their fill and drunk their wine. But the meal had provided food for the mind as well as the body. Kronos had let fall, over sips of the strongest coffee Simplicius had ever tasted, that he had begun to think about a new and radical way of understanding numbers. For reasons he had not yet explained, he had taken to calling his ideas non-Euclidean arithmetic. As the two strolled into the garden, a dialogue began, and it led in such provocative directions that, when the two parted, Simplicius rushed home to write down what he could remember of the exchange. In what follows, Simplicius and I have tried to edit those recollections for clarity, but the words remain, as much as memory can make them, the words of the mathematically minded philosopher and his curious apprentice.

Simplicius. Let's start with the obvious question. What is non-Euclidean arithmetic?

Kronos. That depends on your point of view. Some people might consider it simply an unorthodox new way of thinking about ordinary numbers. Others might look at it and see the death of God.

Simp. The death of God?!

Kron. Or at any rate, the death of the kind of quasi-religious thinking with which mathematicians approach their work. Ask them about the nature of the objects they study—numbers, points, lines, sets, spaces—and you'll get an entire theology. The doctrine is Platonism: the idea that certain ideal objects (in this case, mathematical ones) are "out there" somewhere, existing prior to human beings and their culture, untouched by change, independent of energy and matter, beyond the confines and necessities of space and time, and yet somehow accessible to the minds of mathematicians.

Simp. Do mathematicians really believe that? Or is it just a way of talking that lets them get on with their research?

Kron. They believe it, deeply. In 1996, in the *Times Literary Supplement*, our friend Brian Rotman reviewed *Conversations on Mind, Matter, and Mathematics*, a transcript of discussions between two illustrious French scientists: the mathematician Alain Connes and the biologist Jean-Pierre Changeux. It's a fascinating book. At the outset Connes proclaims his belief in a "raw and immutable reality" (26)—a mathematical realm that exists "independently of the human mind." Mathematics, he says, is the exploration of an "archaic reality"; the mathematician "develops a special sense . . . irreducible to sight, hearing or touch, that enables him to perceive a reality every bit as constraining as physical reality, but one that is far more stable . . . for not being located in space-time" (28). That may be slightly more intense and more poetic than the way other mathematicians would put it, but most of them would agree with Connes.

Simp. How does the biologist react to those ideas?

Kron. He is flabbergasted. How, he demands, can Connes be a materialist and still think like that? How could a physical brain make contact with such an "archaic reality"? But Connes doesn't budge an inch. He happily admits that the "tools devised by mathematicians to understand mathematical reality" are a human invention, formed (or "tainted," as he puts it) by history. But that has no effect on the "raw reality" the tools are used to investigate. In fact, he suggests, the physical world is created out of the mathematical one—an idea that the numerical mystic Pythagoras propounded more than a century before Plato.

Simp. So mathematical Platonism has pretty deep roots!

Kron. Deep but not firm. In fact, I'm convinced its days are numbered. Mathematicians will cling as long as they can to the belief that their art brings them into contact with an eternal world. But that idea can't hold up forever. Not with computers around.

Simp. Computers?

Kron. For two decades computers have been helping create a new kind of mathematics—experimental mathematics. Thanks to the explosion in graphics and imaging software, mathematicians can construct mathematical realities and then manipulate and visually explore them. Now they can produce previously undrawable diagrams such as fractals and chaos maps; they can visualize topological surfaces whose existence was unsuspected before they were seen on a screen; they can discover features that precomputational mathematicians never could have imagined.

Computers are even starting to change what is meant by a mathematical proof. Twenty-one years ago mathematicians were shocked when a computer proved the so-called four-color theorem—the theorem that asserts that no more than four colors are needed to color in all the regions of any map in the plane. The computer had to check millions of details on some two thousand complicated maps that no human mathematician is ever likely to see. Traditionally, giving a proof had always meant giving a convincing logical narrative that a mathematician could follow step by step. The proof of the four-color theorem certainly wasn't *that.*

Now even more advanced methods are coming into play. One of them, known as automated reasoning, made headlines just a year ago when a computer settled a difficult sixty-year-old problem known as the Robbins conjecture. Probabilistic proof procedures have also been invented, in which a computer runs spot checks on a complicated proof to determine how likely the proof is to be correct. The catch—as the name implies—is that probabilistic proof is never totally certain. The use of such methods has led some mathematicians to predict the coming of a "semi-rigorous mathematical culture."

Simp. All very interesting. But why should that threaten the idea of Platonic objectivism?

Kron. For one thing, because it's clear that the computers themselves are not perceiving mathematics; they are constructing it. By giving mathematicians access to results they would never have achieved on their own, computers call into question the idea of a transcendent mathematical realm. They make it harder and harder to insist, as the Platonists do, that the heavenly content of mathematics is somehow divorced from the earthbound methods by which mathematicians investigate it. I would argue that the earthbound realm of mathematics is the only one there is. And if that is the case, mathematicians will have to change the way they think about what they do. They will have to change the way they justify it, formulate it, and do it.

Simp. It seems to me that ideas like this have popped up before.

Kron. Yes. More than a hundred years ago the German mathematician Leopold Kronecker declared that "God made the integers; the rest is the work of Man." He meant that mathematicians can take it for granted that integers exist, but that they must prove the existence of other kinds of numbers (fractions, irrational numbers, and so on).

Early in the twentieth century the Dutch logician Jan Brouwer carried Kronecker's concern about existence several steps further. Brouwer sharply criticized the methods by which mathematicians prove the existence of mathematical objects. One of the most powerful techniques of classical mathematics was (and is) to prove the existence of something by disproving its nonexistence—that is, to assume that it does *not* exist and then show that the assumption leads to a contradiction.

One famous theorem, for instance, attributed to Euclid, proves that the number of primes is infinite by assuming the contrary: namely, that there is some largest prime p. That assumption enables Euclid to prove that there must also be a prime larger than p, a contradiction. The only way to wriggle out of the contradiction, then, is to deny the original assumption, and that means there is no largest prime, proving the theorem.

That's not good enough, Brouwer said. If numbers and other mathematical objects do not exist in some kind of Platonic realm—if, that is, they are constructed—then the only acceptable existence proof must be a recipe for constructing them. Brouwer spent many years developing such constructive existence proofs. Then, in the 1960s, the American mathematician Errett A. Bishop, then at the University of California at San Diego, showed that most of classical mathematical analysis can be proved constructively. The bonus was that constructive proofs tend to be much more informative than traditional proofs are.

Simp. So mainstream mathematicians were glad to see them?

Kron. No. They ignored them. The main reason is that most of twentieth-century mathematics has devoted itself to exploring the properties of infinite sets of whole numbers and real numbers as if they were completed (rather than constructed) entities. Seen in this way, such sets resist the methods of Brouwer and Bishop. Loosely speaking, if you accept constructivism, then you have to abandon the idea of an infinite set.

Simp. Yes, I can see that would be going too far.

Kron. On the contrary, it doesn't *go* far enough—not in the case of *experimental mathematics*. As mathematicians increasingly embrace the computer, the entire theology of "out thereness" will go out the window. And because computers are real objects—at least, they are objects that must be potentially realizable in this universe—mathematics must learn to confront the realities of computation: real time, real storage, real energy, real error, real instructions, and real, material implementation. In other words, it will run up against real-world limits—finite limits. Non-Euclidean arithmetic is an attempt to show how mathematics might cope with those limits. Its key is the concept of integer—one that doesn't go on forever.

Simp. Why do you call it non-Euclidean?

Kron. It's analogous to non-Euclidean geometry. In the classical geometry of Euclid, points and lines are supposed to reside on an infinitely extended, already existent, everywhere identical plane. Their rock-bottom properties are described by simple, seemingly self-evident axioms. The fifth of Euclid's axioms states that through a point external to a given line, one and only one line can be drawn that is parallel to the given line. That seems intuitively obvious if you look at a point and a line in the plane.

But early in the nineteenth century, mathematicians discovered that they could formulate internally consistent systems of geometry by assuming that more than one parallel line, or alternatively, no parallel lines, could pass through the external point. Those so-called non-Euclidean geometries turned out to be ideal for describing lines and points on spheres and other nonplanar surfaces. In fact, as Einstein showed, the geometry of the universe itself, on a cosmic scale, is non-Euclidean, even if on a human scale its divergence from flat, Euclidean geometry is so small as to be unnoticeable. But even if non-Euclidean geometries had turned out to be useless, their mere existence was enough to shatter the idea that the Euclidean plane was some kind of uniquely privileged Platonic realm.

Euclidean *arithmetic*, then, is just familiar, classical arithmetic. I call it Euclidean because it rests on the Platonic idea of numbers as an already existent, infinitely extended series of objects, each different from its neighbor by an identical unit. It treats all numbers, even the impossibly large ones, as if they behaved exactly the way the familiar, local numbers do. But what if arithmetic, like geometry, is only *locally* Euclidean? Why should mathematicians assume that, say, arithmetic operations with extremely large numbers—numbers you can't even write down without special notation—work just the way they do with numbers of more "ordinary" size, numbers you (or a large computer) might be able to compute with in standard decimal or binary notation? Non-Euclidean arithmetic rejects Platonic ideas about numbers by rejecting one key part of the Euclidean arithmetic scheme.

Simp. The concept of infinity.

Kron. Right. Specifically, it does away with the ad infinitum principle that allows one (whoever and whatever "one" is supposed to mean) to go on doing something endlessly. In this case something is counting: the exact identical adjoining of a unit over and over. The basic idea goes back to Aristotle. In his view, infinity means being able to do what you've just done again, and again, and again, and so on. No individual step ever reaches infinity, but the process transcends finitude because potentially it can go on without limit. Non-Euclidean arithmetic says you can't do that.

Simp. But why not? There's always a next number. Say you name a number n. I can come back with $n+1$. How can anyone rationally deny my ability to add one to n and get $n+1$?

Kron. It's true, being able in principle to do something ad infinitum is deeply embedded in Western thought. It appears utterly natural, obvious, undeniable. To see how strange it really is, you have to step aside and ask: Who or what is adding or continuing? Who or what is assigned the task of endless counting? You have to look closely at the language used by mathematicians—how they operate with and respond to signs, and especially how they interpret injunctions and open-ended instructions like "go on adding a unit."

In his book *Ad Infinitum*, Rotman spends several chapters doing just that. Simply put, he concludes that counting is an idealization of certain simple procedures people undertake in the physical world:

laying stones in a row, stringing beads, making marks with a pencil. We mathematicians idealize the procedure, the marks, and the person or machine that does the counting.

Now, in mainstream mathematics, that idealized doer—the Agent, as Rotman puts it—can count ad infinitum, and the process at every step will always be the same. That's because the Agent lives in a frictionless, airless, timeless, Platonic world where it can count as long as it pleases without cost or effort. But suppose you decide not to idealize quite that much. Suppose you decide there is a cost to counting—that as you go along, counting gets increasingly difficult until it stops altogether. In the real world, your calculator batteries might run down. Or your pen might run out of ink. In any case, eventually you will reach a point at which what you are doing has changed so much that it no longer constitutes counting. It really doesn't matter what kind of limit there is, or how big it is. There needn't even be a precise numerical boundary. The important thing is that a boundary is there.

If you then introduce the arithmetic operations of addition, multiplication and so on, in the usual way, it turns out that you can study the kind of arithmetic that takes place under those conditions, and you can even draw up axioms that govern its properties. And when you do, you get an intriguing alternative to the classical view—a more realistic alternative, I would argue.

Simp. What exactly *do* you get?

Kron. You get a number system—a number line, if you will—like the one for Euclidean arithmetic but with some important differences. For example, the line is bifurcated into two kinds of numbers: countable ones and uncountable ones. Within the countable numbers, arithmetic works the way you learned it in elementary school, except that each operation will inevitably specify numbers that can't be reached by counting. For example, if you count upward, you'll reach a countable number x such that x plus x is not countable. Before you get there, you'll have reached a countable number y such that y times y is not countable. And before that, you'll have reached a countable number z such that z to the zth power is not countable. I call the countable numbers *iterates* and the uncountable ones *transiterates*.

Simp. So $x + x$ and $y \times y$ and z^z are off-limits.

Kron. Not exactly. You can still use them to name numbers, and still write the symbols for $10^{10^{10}}$ or any of the staggeringly huge numbers that mathematicians have devised for specialized applications. You just can't reach those numbers by counting. The transiterates are numbers that you can name but not reach.

That distinction gives rise to some unusual properties. For example, suppose you come from a culture in which counting stops when you run out of fingers. In that case, the transiterates start with the number 11. It's clear that addition is commutative when the result is an iterate: three plus seven is the same number as seven plus three. But are transiterates commutative as well? Is four plus seven the same number as seven plus four?

Simp. It would have to be, wouldn't it? That's how numbers behave. Seven plus one is the same as one plus seven. Seven plus two is the same as two plus seven, and so on.

Kron.[*aside*] One, two, three, dot, dot, dot. [*to Simplicius*] But the trouble comes with your "and so on." In both your examples you are iterating, which is exactly what can't go on happening. "Seven plus one" means "count to seven, and count one more." With seven plus four you are dealing with a number our fictional culture can't count to. It's just as inaccessible as 10^{100}. If you like, you can state an axiom requiring the transiterates to be commutative, but nothing forces you to do that. Similarly, you could state an axiom that says that if a is greater than b, and c is greater than d, then $a + c$ is greater than $b + d$. In that case the transiterate $8 + 5$ would be greater than the transiterate $7 + 4$. Setting things up that way would force the transiterates into some kind of order. But it's not necessary to do that either.

In a sense, then, the transiterates aren't really on the number line at all. You might imagine them floating off to the side somewhere, as if the number line fans out into a spray of disordered numbers.

Simp. Whereas within the iterates, normality prevails.

Kron. Not entirely. You still have the rational numbers, or fractions, to worry about. For example, ordinary rational numbers are what mathematicians call *dense*: if you take any two of them, you can always find another one in between. That's easy to see: just take their average.

But which fraction lies between $1/9$ and $1/10$? The only possible candidates have transiterate denominators—denominators greater than 10—and transiterates, as you have just seen, have no fixed address. Similar problems crop up when you try to put fractions in order by size. Is $4/6$ the same fraction as $2/3$? "Obviously," you say. But not when you remember that, by definition, $a/b = c/d$ if and only if $a \times d = c \times b$. In this case, the relevant products, 4×3 and 2×6, are both transiterates, so there's no way of telling whether they are equal. Hence fractions, too, float free of the number line.

If the Euclidean number sequence is an unswerving line, a laser beam shining down a long hall to infinity, I picture the non-Euclidean number sequence as a line of iterates with the transiterates cascading out of them. Or, to evoke another image, it's a circulatory system, narrowing into capillaries as one gets closer to any given rational number, and billowing out at the far end into a transiterate heart. Only in between the extremes are things comfortably normal.

Simp. And all that is supposed to be more realistic than the way we ordinarily think about numbers?

Kron. Well, the number 10 isn't very realistic, of course, but the basic idea is. If numbers are constructed rather than discovered, and if they are rooted in the real universe rather than in some Platonic heaven, then whatever they are made from has got to be a limited resource. So numbers must run out sooner or later.

Simp. But even granting, provisionally, that your iterates stop somewhere, that "somewhere" has to be a lot higher than 10, doesn't it? Where is it, really?

Kron. There are several ways of approaching that question, depending on your assumptions. The English astrophysicist Sir Arthur Stanley Eddington, for instance, attached a kind of numerological significance to the number of particles in the universe, which he calculated to be around 10^{80}. One might think of that number as the "somewhere," which would have a crude intuitive appeal—you stop counting when you run out of things to count. But I think it's possible to be more sophisticated than that.

For example, you could imagine, as we did in Chapter 3, the counting being done by an ideal computer, which occupied a bounded portion of the universe and used the theoretical minimum of energy. Thermodynamics and information theory provide formulas for how much energy such a computer would consume to count up to any given number n. In theory, the computer could keep counting until it had consumed all the energy in the universe. No one knows how much energy that is, but physicists have estimated U, the mass-energy of the *visible* universe, at around 10^{75} joules. On the basis of that figure, an ideal computer could count to about 10^{96} before running out of energy.

Actually, you could do better than that. If you imagine that the machine is as big as the universe, it wouldn't be necessary to expend the energy needed to store all the preceding numbers. The state of the universe itself could serve as the computer's memory, and one would only need to allocate energy to store the current number. In that case, counting would halt when the universe-computer needed an amount of energy equal to U to count one single step, from n to $n+1$. Even under such extravagant conditions, the computer couldn't get beyond $10^{10^{98}}$—which, you could say, is the outer horizon of all counting in this universe.

As I've said before, though, the crucial thing about a limit to counting is not where the limit lies but that it exists. If you tried to count that far from inside the universe, using a real computer with real energy requirements, you would use up more and more of the fabric of the universe trying to get there. Classical arithmetic doesn't reflect that reality. Non-Euclidean arithmetic does.

Simp. So in order to face reality, everyone should scrap the old arithmetic and switch to the new, more realistic one.

Kron. Oh no, not at all, any more than schools should stop teaching Euclidean geometry. I'm merely promoting non-Euclidean arithmetic as another possibility that can be coherently imagined, and that might prove useful.

Simp. Useful for what?

Kron. Potentially, for keeping physics mathematically honest, or for finding an arithmetic appropriate to the materiality of computer-inflected mathematics. That assumes, of course, that the people in those disciplines are interested.

I also think that non-Euclidean numbers might be useful for studying time. Mathematicians have long suspected a link between numbers and time. For example, the nineteenth-century Irish mathematician William Rowan Hamilton confessed: "I do not find myself able to frame a distinct *conception of number*, without some reference to the thought of time. . . . I cannot fancy myself as *counting* any set of things without first ordering them, and treating them as successive" (1853, 15). The idea of a temporal analogue to non-Euclidean space—in terms of spatializing time, of imagining it spread out on a surface—has intrigued me for some years now. The way transiterates fan out from the non-Euclidean number line may provide a way of working it out.

In particular, the distinction between iterates and transiterates may provide a tool for investigating the opposition between parallel and serial actions, between doing lots of things at once instead of one thing after another. I've been fascinated by that problem, as part of a study about the effects of technology on human consciousness.

Simp. Those sound like deep philosophical waters.

Kron. Yes. We are really talking about philosophical questions. In a time of questioning and skepticism about every source of authority, mathematics alone is still widely seen as a font of timeless, indubitable truths. That privileged position was bound to come under fire sooner or later. In that sense, non-Euclidean arithmetic might well be the vanguard of a postmodern mathematics. I can't say where that would lead. But if I'm right that modern mathematical rigor isn't written in stone and that sociocultural reality (in this case, the sequence of numbers) is constructed, then the next step is to start on the social *re*-construction of reality.

Simp. What would you call that? Postconstructivism?

Kron. Enough, enough.

NOMAD INFINITY

Inside contemporary science and technology, the question of infinity—as material effect, concept, affect—specifically the mathematical infinite, flits across the horizon like an almost invisible specter. Much is at stake and much would follow from confronting rather than repressing this disembodied ghost of the endless and the eternal. This is so not only on a general level of mathematics' philosophical and rhetorical self-presentation and physics' still-operative (indeed, uncritically celebrated) Galilean/Pythagorean assumption of the universe's mathematical structure, but, closer to technoscience, within the Platonic metaphysics governing the idea of a computation, a metaphysics introduced into computer science with Turing's replacement of the thinking/sentient computing body by a de-corporealized agent, without whose freedom from all things material his eponymous machine with its endless, potentially infinite tape cannot even be formulated.[1]

And yet, despite repeated refusals of idealist metaphysics, ontotheologies, and transcendentalisms, the question of the infinite (and what is more theological and transcendentally ideal?) remains curiously unthought within contemporary philosophical discourse. In the case of analytic philosophy, which surrendered its critical apparatus to mathematical logic a century ago, this might not seem surprising, since the unchallengeable, God-given obviousness of the endless sequence of "natural" numbers at the heart of mathematics renders any critique of the infinite from within that logic unthinkable. Outside the analytic camp, the situation is more varied but no more satisfactory: one has deconstructionism that doesn't—can't—move beyond the Hegelian good/bad infinite and neither knows nor wants to know anything of a mathematical metaphysics, a Nietzscheanism that rejects such idealism but talks nonetheless, without irony, of *infinite* recurrence and *eternal* return, or an avowedly materialist pragmatics that well understands the importance of number and numbering but seems unable/unwilling to challenge the accepted mathematical presentation of its infinite status.

It is this last, specifically the work of Deleuze and Guattari, that forms the context of the present perspective. True, there is little in their writings to indicate a departure from the transcendentalism inherent in the mathematical community's reading of infinity (either the Aristotelian potential infinite or the Platonist-inspired actual infinite). But they are nonetheless alive, through their discussion of the Nomad, to a view of number opposed to the "royal" or State version, which internalizes these infinites. This, together with their call for a rejection of Aristotelian hylomorphism (precisely the subordination of matter to form necessary for the potential infinite) in the name of a "materiality possessing a *nomos*," justifies our attempt here to modify their enterprise to accommodate a de-transcendentalized—corporeal, material—account of infinity. One can go directly to the point, then, and put to them an appropriately tailored question concerning number.

If nomads were to count, then what kind of number would they produce? Could they mimic the endless counting of royal mathematics and invent their own form of infinity? Surely, this would contradict what is said of them and nomad science and nomad number in *A Thousand Plateaus*. Is not the "Treatise on Nomadology"—the nomad, nomad science, the nomad war machine—structured by the opposition between a State that appropriates nomad inventions and a nomad that copies nothing but intrinsically invents all that it has and does? And yet, might it not be possible and maybe even profitable to interpret the opposition of State and Nomad as less absolute; to still insist on the nomadic refusal of the State, its implacable exteriority from it, and yet imagine the nomad of the twenty-first century as *moving on* to a reverse capture, a nomadic appropriation of the arithmetic machine of the State? In what follows, I want to sketch one particular working out of what a revisioned arithmetic or nomad science of number and infinity could be. Such an arithmetic will perforce depart from the letter and perhaps even the spirit of nomadism as Deleuze and Guattari elaborate it, and in this will result in what might be called an *itinerated infinity*. Specifically, what is proposed here is to think counting as material movement, as a performed activity, and infinity as the arbitrary prolongation of that movement. The resulting divergence from the standard idea of counting will be the manifestation on the level of theory of a prior and elemental departure, a swerve or *clinamen* whose effect—there from the very moment one begins to count—will be the uneliminable presence within the act of counting of the material universe itself. In this, what is offered is an extension or addition, rather than a violation of Deleuze and Guattari's pragmatics, a development of number outside what I call the Euclidean State along a path of materiality and physical performativity entirely consonant with the ethos of *A Thousand Plateaus*'s embrace of nomadism.

DELEUZE AND GUATTARI AND THE NOMAD

Let's note first the two principal relevant oppositions, within number and science, giving rise to different kinds of number and science, respectively, which Deleuze and Guattari understand as operating between the State and the Nomad.

Two Kinds of Number

Number makes its appearance in the writings of Deleuze and Guattari in *A Thousand Plateaus* within the Treatise on Nomadology as one of the three principal aspects of the war machine: "Axiom II. The war machine is the invention of the nomads (insofar as it is exterior to the State apparatus and distinct from the military institution). As such, the war machine has three aspects, a spatiogeographic aspect, an arithmetic or algebraic aspect, and an affective aspect" (1987, 380).[2] The exteriority of the war machine from the State and its military apparatus is reflected in a difference in kind between two sorts of number—the *numbering number* of the Nomad, and the *numbered number* of the State—a difference that lines up with the major bifurcations that form the rhetorical and discursive armature of *A Thousand Plateaus*. Thus, the numbered number, essential to the State's military apparatus, is associated with striated space, signifying regimes of signs, major science, the logos, and the passage-complex of iteration-history-reproduction, while the numbering number, an axiomatic component of the war machine, is associated with smooth space, countersignifying regimes of signs, minor science, the *nomos*, and the movement-complex of itineration-geography following.

In ordinary parlance, which is of course that of the State, the numbered number is the omnipresent and undifferentiated idea of "number," and its significance for the State reaches far beyond purely military uses. "Arithmetic, the number, has always had a decisive role in the State apparatus" (389), not only for the needs of an original "imperial bureaucracy" but also for further instrumental purposes within all subsequent forms of the State, from the management of people to the technoscientific control of every aspect of material reality. The numbered number is the cognitive basis of the State's power, the number on which all measurement, calculation, and technology runs. State or royal mathematics has formalized this number into the concept of integer (together with its derivatives of rational, real, complex numbers, and so on), which, within the imperial monism of Western reason, the State calls simply "number." The numbered number thus forms an all-purpose measurement machine controlling vast quantities that, by their very acquiescence or subordination to the process of being so numbered, are abstract, disembodied, and detached from any concretely given, operating, or manifest pluralities.

Against this, Deleuze and Guattari set up the numbering number, "autonomous arithmetic organization," dealing in small quantities and specific ensembles. They detail two characteristics of the numbering number in its role as aspect of the nomad war machine. The first has it as "always complex, that is, articulated" (391), by which they mean it occurs within specific assemblages—such as the 1 × 1 × 1 of man-horse-bow—and thereby has an internal heterogeneity related to "several bases at the same time." Generally, then, numbering numbers have an internal—protoalgebraic—relational structure. Such a number is not *abstracted* from the arrangements, assemblages, and processes wherein it occurs and is not detached as a homogeneous and interchangeable entity—an abstract "3" instead of the three inseparable "units" of the example here. The numbering number is encountered and understood in situ, embodied or incorporated in that which it is said to number; it is, in other words, number *in performance of itself*. The second characteristic arises from the fact that the war machine "displays a curious process of arithmetic replication or doubling" (391), which gives rise to a certain form of transverse communication. Men are shuffled and organized into a force independently of their lineages, while at the same time individuals are selected from each lineage to form a separate and special "numerical body" that performs certain essential functions within the war machine. What Deleuze and Guattari describe here as a special body with properties over and above those of its members is the war machine correlate (although they don't speak in these terms) of the algebraic concept of a quotient structure.[3] Together, these two characteristics suggest that algebra, no less than the war machine itself, might be thought of as a nomad invention (more precisely, that algebra is the product of the same abstract diagram which nomadism concretizes). This feature, rather than the purely numerical nature of the numbered number, is further layered in Deleuze and Guattari's discussion of what they call a countersignifying semiotic (regime of signs corresponding to animal-raising nomads as opposed to hunter nomads), where it appears as a "numerical sign . . . not produced by something outside the system of marking it institutes,"[4] and which consequently determines "functions and relations . . . arrangements . . . rather, than totals, distributions" (118).

Deleuze and Guattari relate this opposition, the intrinsicality of the numbering number against the extrinsicality of the numbered number, to Boulez's distinction in music between two sorts of occupancy, namely, smooth space-time, as that which is "occupied without being counted" and striated space-time as that which "is counted in order to be occupied." With the difference that they talk of "space" where he talks in terms of "space-time," Deleuze and Guattari use Boulez's distinction in a strategic way throughout *A Thousand Plateaus*. Thus, they see striated space (such as the coordinatized Cartesian plane) as always already measured, divided up and gridded by a counting apparatus that must be activated in order for anything to enter it, and as abstractly homogeneous, exterior, and prior to what is numerated within it; that which achieves this numeration being precisely the numbered number. On the other hand, smooth space, for example, a nonmetricizable or only locally metricizable topological space, according to a geometrical analogy with Boulez's formula, would never allow number (strictly speaking: counted, ordinal number) to be in any way detached from the act and process of whatever comes to occupy it. In other words, in one sense of numbering number, it is not abstractable, it cannot be presented as a "pure" plurality, although, as we shall see, it must encounter something other than itself by the very process of its self-actualization via counting.

Two Kinds of Science

Deleuze and Guattari introduce their conception of nomad or minor science along the outlines (as laid out by Michel Serres) of the atomism of Lucretius. Such a science is a hydraulic model of thought, foregrounding flux and flow over solidity and permanence; it is a science of difference in which becoming and heterogeneity replace identity, sameness, and eternal truth, and it gives a crucial conceptual/explanatory role to the celebrated *clinamen*. The opposition of flux and identity, flow and permanence, nomadic and State science, lines up here with *nomos*, signaling an internally produced model of exploration and theorizing from below and logos, indicating a legal, synoptically imposed from above, model. (In this sense the opposition is reflected *within* certain forms of contemporary State science: so-called cognitivism and conventional robotics which understand intelligence, as we saw in Chapter 4, as top-down, controlled by a pregiven plan or "mind" from above as against various models of "situated" intelligence and subsumption architectures in robotics that insist on bottom-up theories and constructions.) Thus, logos understands science as the finding of preexisting laws, which hylomorphically "govern" matter, subordinating content to a controlling form, with respect to which matter's singularities are divergencies or exceptions within a space striated by parallel layers, identical inescapable verticals, a "space of pillars." Against this, *nomos* names a science that escapes hylomorphism in favor of local, matter-determined regularities that "seize or determine singularities . . . instead of constituting a general form" (369) and which operates in the smooth space determined by the *clinamen*, the swerve or minimum difference that creates, from the outset, a divergence between vertical parallels and the curvilinear—a preemption and creation to which I shall return below.

Finally, and as it turns out most important for the project of understanding what it means to construct a form of nomad counting, there is the radical separation between the passage-complex of iteration-history-reproduction that operates within State science and itineration-geography-following, the movement-complex characteristic of nomad science. Deleuze and Guattari introduce this in terms of the opposition between "following" and "reproducing" that corresponds to a distinction between two forms of scientific procedure. In a science predicated on reproduction, the operative logic is one of seeking repetition and replicability, on the production and discovery of the same, on iteration and reiteration. The goal is always to find invariants of phenomena: to extract from the manifest differences over time and space that which is constant and reproducible, and so arrive at the underlying "law" of material phenomena; a law prior to and outside that which it "governs." "Reproducing implies the permanence of a fixed point of view that is external to what is reproduced: watching the flow from the bank" (372). Reproduction, then, is the method of science interpreted as logos. On the other hand, in a science predicated on "following," one is always within the river, with no access to (or indeed conceptualization of) a bank from which an external understanding is available. The operative logic is that of immersion, a situatedness within the ground or context of the object under investigation, and the goal can never be other than the creation of an endogenous or intrinsic scientific knowledge. Such knowledge arises from following the directions that singularities in matter impose and that—far from being subsequently theorized as falling short of a prior ideal, as exceptions to a preexisting law being discovered—constitute the vector along which knowledge itself is constructed. And as the State counterposes history as its passage through time to nomad geography and movement through space, so their respective sciences replicate this division. The science that reproduces the same in the form of an absolutely enforced temporal repetition—identical iteration—is counterposed to the science that follows the path of difference and movement—performed itineration—in space.

NOMAD COUNTING

We now need to reconstruct nomad number and nomad science and extract from them a practice of counting and arithmetic suitable to a twenty-first-century Nomad cognizant of royal mathematics and willing to appropriate it on its own terms. Such a practice, although apparently similar to and initially indistinguishable from the State version, is ultimately in radical opposition to it.

In royal or State mathematics, the idea of counting is familiar and unproblematic. One starts from 0, and then passes to 1, 2, 3, 4, and so on, where the understanding is that the numbers are construable as bunches of iterated units. There is no sense that counting creates or produces these numbers that, on the contrary, already exist and are merely being enumerated or assigned a series of numerals from some system such as the familiar Hindu-Arabic one. In terms of Deleuze and Guattari's metaphor, these numbered numbers are floating downriver off to infinity, viewed from a riverbank external to them. The goal of the project here is to get into the river and reconstruct the process from there. What obstacles might the attempt to appropriate royal counting present to a nomad understanding of number, and could they be overcome within the boundaries of what we have just summarized as nomad science? The principles controlling the answer to this question are already in place. Instead of an already numbered number in striated space, one has a numbering number associated with a smooth space, and the fundamental move of counting—doing one thing, then another—must operate within the regime of itineration-following and not iteration-reproducing.

Royal counting counts what preexists it; it takes place in a striated space that has already been numbered before anything enters it. The prior striation of the space is unavoidable, and its circular relation to counting invisible: one uses royal counting to define the space in which such counting "subsequently" takes place. To count within a smooth space, on the other hand, is to occupy space without any prior act of counting. Counting in such a space is constitutive: the work of counting is that which brings the space into being along with the numbers which, at the same time, populate it. Royal counting—0, 1, 2, 3, . . .—formal, going on forever, endlessly the same: the ideogram ". . ." being an order-word meaning "go on repeating without stopping." To switch from this royal (immortal and hereditary) body of mathematics to nomad counting—from pure iteration of a pregiven same to an itineration that produces a difference—is to move from a transcendental ideality to an ideal materiality that goes as far and lasts as long as it manages (in its materiality) to do so. Unlike an iteration, which can be reified and treated as having already taken place without ever needing to be executed, an itinerary is nothing in advance of itself; it doesn't exist until it is effected, until it is performed, enacted, and carried out by a real or imagined or virtual movement of bodies. The steps or stages of an itinerary, unlike those of an iteration, are not there, present, waiting somewhere in advance of their effectuation: one performs a step, then another, then another, where each "then" comes into existence only after the previous one, by virtue of an embodied movement, has itself been realized. If royal counting furthers itself by a necessarily unchanging reproduction, nomad counting does not *re-produce* anything; it *produces* itself, one step at a time, by following, in the sense of being always one step behind, its own material process.

Nomad counting, then, is embodied, performed itineration. But, granting that such counting is possible in contradistinction to the disembodied, iterative counting of royal mathematics, is there not a doubt, in relation to its putatively nomad character, about the kind of *number* it produces? Such a number, insofar as it is "just a number" (autonomous, prolongable, and detached from any concrete situation) is certainly not confined to small quantities and specific ensembles and seems removed from the in situ idea of a numbering number explicated earlier (for example, the triple 1 × 1 × 1 and not 3), which possesses an inner heterogeneity, an internal structure, differentiating it from a pure, de-territorialized magnitude. To respond to this objection, a number of considerations are in order.

First, it is necessary to point out that number, if it is to take part in any sort of arithmetic, is never not abstract. Which means that precisely in order to provide the wherewithal for a science of number, even nomad counting is understood here as a pure, limit form for the production of number; number as maximally unencumbered, decoded, and de-territorialized as possible, wherein the most free and minimal interpretation of any demand for situation, concreteness, and inner structure operates. In other words, nomad numbering turns the materiality that royal numbering refuses to recognize, namely, its own active presence, its complicity in the creation of numbers, into a positivity, into its own becoming; a becoming that the dream of the State—pure and absolute de-territorialization—invisibilizes into the preexistence of number that was never "created."

Second, in relation to this, one needs to distinguish between cardinal and ordinal numbers, since while it is true that cardinals need not have inner structure (other than their presentation as arithmetical combinations of ordinals or of other cardinals) and can appear solely in the guise of pure aggregates or instantaneously presented magnitudes, this can never be the case with ordinals. The ordinal 3, for example, as opposed to the cardinal 3, has within it the structure of its past—first, second—actions; any ordinal contains the path to itself, that is, the means by which it comes about *as an ordinal.* Although the distinction is systematically blurred within the formalism of everyday arithmetic, it is clearly made and foregrounded within the more comprehensive context of abstract set theory, where a definition of ordinals is given that specifies them as certain kinds of structured sets; thus each ordinal number is defined, after von Neumann, to be the set whose members consist precisely of all previous ordinals: 0 is the empty set $\emptyset$; 1 is the set $\{\emptyset\}$, whose sole member is the empty set; 2 is $\{\emptyset, \{\emptyset\}\}$; and so on. Ordinals, then, which are the numbers produced by nomad itineration, no less than by classical iteration, have an irreducible "non-numerical" structure, which means that although they certainly extend it to the limit, itinerated numbers do not violate the demand framed for numbering numbers that they be inhomogeneous and internally differentiated.

Third, on more than one occasion, Deleuze and Guattari's own gloss on the numbering number, especially in the context of the connection between such number and smooth space—as in the description of it as "rhythmic not harmonic" (390) and "ordinal, directional" (485), and so on—suggests an interpretation of nomad number in terms of serial continuation (rhythm, ordinality, melody) and not parallel presentation (cardinality, harmony) and hence, although they don't draw it out thus, legitimates seeing the numbering number in terms of ordinality and the ineradicable business of counting.

Finally, observe that the divergence between nomad number, as it results from counting, and the strictly limited, protoalgebraic and small-quantitied numbering number of their original characterization matches the difference between what Deleuze and Guattari call the "two poles of the war machine." At one pole, the war machine has as its object war itself: it then "forms a line of destruction prolongable to the limits of the universe" (422); the other pole, when the war machine operates with "infinitely lower 'quantities,'" it has as its object not war "but the drawing of a creative line of flight" (422). Nomad counting, formulated as a reverse capture of State counting, belongs to the first pole. It gives rise to a nomad infinity, which, as we shall see momentarily, can be interpreted as a form of materializable counting that is precisely and nothing other than a prolongation of number to the "limits of the universe."[5]

Evidently, State and nomad counting give rise, when arbitrarily prolonged, to two kinds of infinite progressions: the infinity of classical mathematics synonymous with the endlessness of the so-called natural numbers; and unbounded itineration, the continued movement along a line drawn from point to point by this very movement of the Nomad. The first, top-down, from a transcendentally external viewpoint; the second, bottom-up, from an always prior and intrinsic materiality. Can one say more? We started from the idea of a nomad appropriation or mimicry of the progression of natural numbers on which the arithmetic of the State mathematics is founded. On this basis, one would expect a nomad arithmetic to emerge. And this is indeed what happens, as we shall now see.

THE NON-EUCLIDEAN SWERVE

We can short-circuit the particular complications that Deleuze and Guattari's treatment of nomadism might introduce into the question of nomad arithmetic through the following observation. The question of materiality is primary to nomad counting and its passage to the limit in comparison to the royal versions of these concepts. Royal mathematics understands whole numbers and their arithmetic limit hylomorphically; this is true whatever philosophical account it gives of itself (Platonist, formalist, constructivist), since all such accounts agree with the first half of Kronecker's dictum that God made the integers: the integers are originless, given, already there independent of any mathematical work, however mathematicians might quarrel about where and how the rest of mathematics arises. The immediate implication of this understanding is that individual numbers appear as ideal "forms" necessarily prior to the material "instances" and "examples," which are supposed to illustrate them and constitute their content. Moreover, the field of these material instantiations (suitably idealized so they can enter mathematical discourse) is what royal mathematics means by finitude; and it suffers another, internal version of hylomorphic subordination, in being thought as that which falls short of the prior ideal form of infinity. Nomad mathematics denies both halves of this hylomorphism: the whole of its arithmetic is immanent in the ineluctable vicissitudes and opportunities of a materiality embodied and operative through itineration. Numbers and their passage to the limit we are ascribing to them here exist only through the performance of counting, immediately so as individual entities, and more widely so through the possible arithmetico-geometrical constructions the assumption of this counting and its limit facilitates. What constitutes their "form" as abstract objects follows from this determination. Thus, what can be said in general mathematical terms about nomad numeration, namely, its arithmetic, becomes a special case of the arithmetic(s) that emerge from embodied or physicalized counting *as such*.

In the past few years, I have investigated the possible origin and necessity of such arithmetics within the Peircean framework of thought experiments explicated in Chapter 1 (Rotman 1993a).[6] To complete the picture of nomad arithmetic, then, I shall summarize what this investigation reveals, and then link it specifically to nomad science, nomad infinity, and the idea of the *clinamen* mentioned earlier.

Within this investigation, what I am calling nomad counting is designated as *non-Euclidean* (Rotman 1993a).[7] It is opposed to the Euclidean kind, what I've been calling *royal counting*, which, like the Euclidean plane, is homogeneous, uniform, pregiven, and always the same; the paradigm in fact of the purest numerical striation. Thus, non-Euclidean counting is the counting of an imagined but always (theoretically and in principle) materializable/embodiable counting agent. Initially, then, the numbers it produces, counting from 0, then 1, then 2, and so on, are indistinguishable from the 0, 1, 2, and so on, of royal counting, but they are not *identical* to them. As each kind of counting is prolonged, the arithmetic properties of the numbers associated with it diverge. Whereas Euclidean numbers, however large, exhibit precisely the same local structure—nothing of mathematical significance distinguishes the action of adding a unit to an arbitrary n to obtain $n+1$ from adding a unit to 1 to obtain 2—

non-Euclidean numbers exhibit structural differences related to their ordinal magnitude: so that, for example, for large enough numbers n, the sum $n+n$ is a number representing a magnitude that spills beyond what is accessible through an act of counting from 0; it is a transcountable number that cannot be conceptualized as an ordinal. Before one reaches such numbers, the same phenomenon occurs with multiplication; there will be numbers m such that $m \times m$ likewise represents a magnitude not countable from 0.

The picture as we observed earlier, then, is of two kinds of number: ordinals countable from zero; and transordinals or cardinals, which, although nameable as arithmetic combinations of countable ordinals, are not themselves countable. This bifurcation of the domain of non-Euclidean numbers into countable and uncountable gives rise to an arithmetic of integers and rationals and hence a number "line" radically different from the classical, Euclidean continuum. While the latter is invariant under magnification or enlargement (that is, any section has a structure identical to any other), this is not so for its non-Euclidean analogue, whose structure in the large is determined by the presence of uncountable numbers and whose microstructure reflects this presence through nonstandard or infinitesimal "rationals" formed from ratios of uncountable numbers whose properties differ from standard rationals. The non-Euclidean "line," then, is ultimately—by which I mean in terms of increasing magnification—nonlinear: the finer the resolution with which the non-Euclidean "line" is examined, the greater the falling away from an unswerving linearity.

The upshot of this, in terms of the use of mathematics to order and measure a physical universe, is that non-Euclidean measurement of time and space manifests a symmetry between the very large and very small; the microcosm and macrocosm, the subquantum world down to the limit of physical subdivision and the entire spatiotemporal expanse to the boundaries of the universe, are indissolubly linked. The reason for this is simple: the behavior of large numbers, and with them their non-Euclidean properties, is transferred through the operation of taking reciprocals to numbers in the small.[8] Numerically, that is, insofar as it is measured and represented by numbers, the entire cosmos sits inside every atom: each instant and each position that are numbered by actual numbers contain virtual subinstants and sublocations—temporal and spatial realities whose arithmetical descriptions reflect the global properties of the measured universe at large.

The situation can be compared, in some respects, to the one Epicurus describes: "In one moment of time perceived by us, that is, while one word is being uttered, many times are, lurking which reason understands to be there . . ." (quoted in Deleuze 1990, 274). Where, for Epicurus's time "perceived," one must understand time measured, for his "reason," arithmetical consequence, and where numerically specifiable positions in space no less than moments of time are also guaranteed, by the same non-Euclidean apparatus, to be lurking in the neighborhood of every actual measurement.

One can deepen the connection between Lucretian atomism and non-Euclidean arithmetic through the notion of the *clinamen* as a minimum declination or swerve from pure, unchanging linearity. Counting starts from the passage from 0 to 1. And already, in this initial step, the difference between the embodied and the disembodied is set in motion. Let us notate the distinction and write 1^e for the embodied unit and 1 for the familiar disembodied unit. Evidently, we are denying the equation $1 = 1^e$. But what does it mean to say that 1^e is not identical and interchangeable with 1? How can there be two versions of the unit, of the singular? The answer is, of course, that they give rise to different kinds of number, that the difference between 1 and 1^e is revealed only retrospectively, under repeated self-adjunction in the presence of the arithmetical operations, and cannot be knowable as an autonomous distinction in advance.

Observe that the presence of arithmetic operations here is absolutely necessary for this effect; repetition or pure self-adjunction—a naked iteration—alone will not do it. Thus, only after + is introduced can embodied counting differentiate itself from the classical kind. Before that, the sequence "0, 1^e, 2^e, . . ." offers no intrinsic characteristics, ones specifiable from within the river as it were, that could separate it from the sequence "0, 1, 2 . . ." After that, one has actual numbers n, which, when added to themselves, result in virtual numbers $n+n$; likewise for numbers multiplied by themselves, and so on.

The situation within the sequence 1^e, 2^e, . . . , is precisely analogous to that described by Deleuze and Guattari for a true abstract machine that "has no way of making a distinction within itself between a plane of expression and a plane of content because it draws a single plane of consistency which in turn formalizes content and expression . . ." (Deleuze and Guattari 1987, 141). Here, the plane of consistency is the naked sequence of ordinals produced by the machine of embodied counting. And it can formalize any content outside itself only through the intervention of a plane of arithmetical expression—the expressive symbolic apparatus of arithmetical operations—at which point the machine ceases to be abstract and becomes distinguishable, by internal, arithmetical means, from disembodied counting. In effect, the entire non-Euclidean conception of number and its divergence from the classical, Euclidean one, the bifurcation into countable/uncountable, and the subsequent infinitesimal quantities designating moments and positions, which, unnoticed and ubiquitous, lurk, as Epicurus attractively puts it, can be understood as the working out of this pregiven, irreducible, and invisible difference between 1 and 1^e.

The *clinamen*, which produces the invisible swerve of non-Euclidean counting and its number line away from the laser beam straightness of classical infinity, is this difference. Or, more specifically, one particular mathematical understanding of it according to which it is the effect, the necessary presence, of materiality itself, as if the entire physical universe were operating inside the arithmetical unit, folded into the arithmetically minimal and semiotically irreducible act of making a mark. In this sense, the *clinamen* is not conceptualizable as an infinitesimal or differential quantity akin to the quantities bearing those names constructed within classical mathematics,[9] since it is precisely that which allows the emergence of a mathematics that refuses such constructions. It is rather the invisible difference between an action and its repetition, and as such is the condition for the possibility of any arithmetic that is able to incorporate (literally) this difference into its idea of counting. Or, to express the point in almost Heraclitean terms, we are dealing here with the difference between performing an action in the universe and "repeating" it, that is, performing it "again" in a universe already changed by the "first" occurrence. Whereas classical arithmetic can assign no sense to the quote marks here, non-Euclidean arithmetic erases them by internalizing the distinctions they notate.

What, then, is the value—the use—of this nomad arithmetic and its infinity? Are there situations where a material infinitude/finitude opposing itself to the disembodied infinity of the State can be applied? One can mean two sorts of application: practical (technoscientific conceptualization of number), where nomad or non-Euclidean arithmetic might matter, and philosophical. The first, in the form of the ultimate limits of computation, was the subject of Chapter 3 and, differently, again in relation to a new model of measured time was broached earlier in the present chapter. On the second, we've already seen how the existence of an alternative arithmetic formalism allows one to recognize the sense in which State or classical understandings of the integers and their passage to the limit is inherently hylomorphic. Deleuze and Guattari urge a model that is "less a matter submitted to laws than a materiality possessing a *nomos*" (408). Evidently, it is royal mathematics that operates within a hylomorphic model, imposing on all matter a prior numerical form, submitting all multiplicities to the law of a pregiven number, and it is nomad, non-Euclidean mathematics that provides materiality, in its appearance as discrete pluralities, with a *nomos*. What else is allowing the materiality of counting, its embodied performance, to determine our conception of number than a "surrendering to the wood"? Of course, it is a maximally idealized wood consisting of no more than a space cleared of all external particularities and reduced to a pure resistance to movement, to an ineradicable friction through which all itineration makes its way. But this is not to say the wood is without singularities. True, no such make their appearance as long as only naked ordinals are involved, but once addition, multiplication, and so on, are introduced singularities become evident: each operation giving rise to a threshold within the progression of ordinals beyond which it ceases to produce countable results.

One can add to this another layer. By ignoring concretizations and obliterating all gradations of finitude as so many fallings short, initial sections, or truncations of an already there infinite, State mathematics is inherently de-subjectified and affectless. The real depression of the spirit, one might say paradox of royal infinity (which Pascal already understood and found so chillingly empty), is not some logical puzzle about the set of all sets that the doctors of mathematical rigor have fussed over, but State mathematics' contempt for finitude and elimination of any possibility of conceptualizing a small/large difference; that is, imagining a comparative understanding of large and small numbering that might engage with life, death, and any human and nonhuman ecological sense of being or having been here and living into the future. Without this dismissal of the "merely" finite, royal infinity loses a key element of its rhetorical purchase, namely, the magical allure of eternal truth and its magisterial negation of anything less than eternal finitude. Only because it provides a system of mythic ideation whereby 10 or 10 billion (or 10^n, where n is itself 10 billion, or indeed any number one can name or imagine a name for) not only is the result of an unproblematic extension of elementary counting but also is as negligible as the unit when compared to infinity, can royal infinity's lure be maintained. But the price of this vision *sub species aeternitatis* is a stultified and diminished relation to numbering that effectively eliminates all but a false finitude, one boxed in and guaranteed by a prior and infinite certitude. What nomad infinity offers is not any kind of certainty or absolute necessity but a "return to finitude and precariousness," to contingency and unpredictable novelty and the positivity of a "lack," a return which, as Guattari so rightly insists, offers a "way out of eternal and mortifying dreams" and hence "gives back some infinity to a world which threatened to smother it" (1995, 96). Only by refusing to sever counting from the material vicissitudes of that-which-counts and building a nomad arithmetic of the infinite on the resulting number is such a return gift possible.

REFERENCE MATTER

NOTES

CHAPTER ONE

An earlier version of this chapter appeared in *Semiotica* 72, nos. 1–2 (1988): 1–35.

CHAPTER TWO

An earlier version of this chapter appeared as "Thinking Dia-Grams" in *South Atlantic Quarterly* 294, no. 2 (1995): 389–415. It was written during the period when I was supported by a fellowship from the National Endowment for Humanities.

CHAPTER THREE

A portion of this chapter originally appeared in *Minds and Machines* 6 (1996): 229–38.

CHAPTER FOUR

An earlier version of this chapter appeared as "Exuberant Materiality" in *Configurations* 2, no. 2 (1994): 257–74.

1. For example, the following statement by Gardner, "Penrose finds it incomprehensible (as do I) that anyone could suppose that this exotic structure [the Mandelbroit set] is not as much 'out there' as Mount Everest is, subject to exploration in the way a jungle is explored" (quoted in Roger Penrose, *The Emperor's New Mind* [New York: Oxford University Press,

1989], 26–28), which is an extended apology and defense of Platonism by a distinguished mathematician. I prefer to quote Gardner (rather than Penrose himself or some other professional mathematician) precisely because he writes for such a general audience.

2. See, for example, W. S. McCulloch, *Embodiments of Mind* (Cambridge, Mass.: MIT Press, 1965); Gerald M. Edelman and V. B. Mountcastle, *The Mindful Brain* (Cambridge, Mass.: MIT Press, 1978); Francisco J. Varela, Evan Thompson, and Eleanor Rosch, *The Embodied Mind: Cognitive Science and Human Experience* (Cambridge, Mass.: MIT Press, 1991); J.-P. Changeux, *Neuronal Man: The Biology of Mind* (Oxford: Oxford University Press, 1986); Mark Johnson, *The Body in the Mind: The Bodily Basis of Meaning, Imagination, and Reason* (Chicago: University of Chicago Press, 1987); Gerald M. Edelman, *Bright Air, Brilliant Fire: On the Matter of the Mind* (New York: Basic Books, 1992). Of course, the agendas, themes, methodologies, and intellectual concerns vary enormously among these works. Thus, to take a single example, Varela, Thompson, and Rosch's *Embodied Mind* invokes Merleau-Ponty's phenomenology of perception as well as embraces a particular Buddhist-inspired idea of the (non)self, whereas Edelman in *Bright Air, Brilliant Fire* (and elsewhere) makes a point of distancing himself from every philosophical position he can think of and would, one imagines, be very uncomfortable in the same bed with Buddha and Merleau-Ponty. And so on. Nonetheless, for the purposes of the broad-brushed picture here, and in relation to the cognitive focus they all have, the juxtaposition of these works is more than justified.

3. For the citation of this example as paradigmatic of what he means by embodiment, and likewise the following one of situatedness, see Brooks (1991b).

4. For a useful and interesting set of responses to some of Brown's ideas, see the debate "Can Research Reinvent the Corporation?" *Harvard Business Review* 69 (1991): 164–75.

5. Readers interested in a more detailed account of the issues behind the anticognitivism sketched here should consult the replies by Lucy Suchman, Philip Agre, and others in the same volume of the journal *Cognitive Science* to a stand-up defense of it by Vera and Simon (1993). Differently, and on a broader cultural terrain than that under observation here, the move being summarized maps onto the refusal of the idea of a controlling, overarching "story" and the embrace of locally determined contingency as an irreducible effect—thus, among others, the loss of what he calls "master narratives" in Lyotard (1984); the insistence on contingent evolution in Gould (1989); and

the argument for the essential contingency of human consciousness in Layzer (1990).

CHAPTER FIVE

An earlier version of a portion of this chapter appeared as "The Truth About Counting" in *The Sciences*, Nov.–Dec. 1997. Courtesy of *The Sciences*, 2 East 63d Street, New York, NY 10021.

1. Less obviously, confronting this ghost has implications for the forms of subjectivity that are being and can be installed by the distributed psyche and for the very idea of immortality at the center of technology's salvationist project. On subjectivity: one has the ongoing reconceptualization of the psyche in which according to Jonathan Crary, a self, a "private unitary consciousness detached from any active relation with an exterior" and inseparable from a "metaphysic of interiority" (1988, 33), is broken open and distributed; a process analogized within a visual metaphorics by the demise of a (Euclidean, dimensionless, transcendental) *point* of view and the formation of an exogenous, postperspectival visual self assembled from material pixels. For an elaboration of this, see Rotman "Becoming Beside Oneself" (forthcoming). On immortality: one has an external refusal of technology's kingdom-of-heaven-on-earth transcendentalism with its Platonic fantasies of the death of death via up/downloading a necessarily de-corporealized mind into silicon.

2. Unless stated otherwise, page numbers in the text refer to the English edition of *A Thousand Plateaus*.

3. In set-theoretical terms, a quotient structure of an algebraic entity A is constructed by partitioning the underlying set of A into a suitably defined family B of nonoverlapping subsets called *cosets*. An arbitrary representative element is chosen from each coset to form the underlying set of the quotient algebra A/B.

4. By which is meant that it doesn't "refer" to anything external, rather than it is not contingent on various larger systems or regimes of the face, of the mask, and so forth, which it clearly is. I shall return to this point below in the discussion of the intrinsicality of nomad counting.

5. One can give a perfectly literal meaning to such a prolongation by setting up a thought experiment wherein a maximally idealized computer is given an input of 0 and instructed to go on counting in (say) binary notation. It is not difficult, as we saw in Chapter 3, to write down the thermo-

dynamic equations describing such a process and read off from them the outer horizon to any such counting within the universe that according to physics we occupy. See the articles cited in note 9 below for some further remarks on this.

6. For the passage to such arithmetics, see Rotman (1993a), where material differences between numbers emerge through an insistence on respecting their construction within three interconnected and simultaneously operative regimes of embodiment: history and culture; the monadic "privacy" of a mathematical subject reading/writing signifiers; imagination, namely, the body of the subject's proxy (what Deleuze and Guattari [1994] call a "conceptual persona") that is imagined to perform thought experiments. It is this last body whose annihilation is demanded by the metaphysical presuppositions of royal mathematics.

7. See especially chapters 5 and 6 in Rotman (1993a), where the idea behind the choice of "non-Euclidean" is elaborated. There is, however, a methodological and consequent terminological difference between what is said there and the present account, which needs comment. Here, I distinguish between itineration and iteration as that which separates the embodied and performative nature of nomad journeying from classical, transcendental repetition. In *Ad Infinitum*, I motivated the treatment of materiality differently by working the analogy with non-Euclidean space. Specifically, I argued for a temporal analogue of the Euclidean/non-Euclidean split and talked of two kinds of iteration—disembodied and embodied—giving rise to classical Euclidean and non-Euclidean arithmetic, respectively. Thus, what are called *ordinals* here, the numbers counted from 0, are designated as *iterates*, and what are called *transordinals* or *cardinals* here are called *transiterates* there. This means that the Deleuze and Guattari opposition between nomad geography (itineration) and State history (iteration) is mapped there onto two models of time.

8. Of course, reciprocals operate in just the same way in the classical, Euclidean case: as n goes to infinity $1/n$ goes to zero. But since there are no structural differences, no discontinuities within classical arithmetic, no novelty is transferred, hence nothing would be asserted by the claim that numbers in the large and the small image each other.

9. Which is how Michel Serres appears to conceptualize the matter in *La naissance de la physique dans le texte de Lucrèce* and how he is understood to do so by Deleuze and Guattari in their citing of Serres's description of a minor science (1987, 361). However, one can try to retrieve Serres's picture: as soon as one reterritorializes the idea of infinitesimal from classical to non-Euclidean mathematics—for example, by referring to the reciprocal

of an uncountable cardinal as measuring an infinitesimal moment or position—then it might make sense to try to "internalize the *clinamen*," to locate it within the very arithmetic it founds, by identifying it in some way as an infinitesimal movement. But, in the context of explicating the *clinamen*'s role as a condition for the possibility of non-Euclidean arithmetic, this idea is unhelpful.

WORKS CITED

Berry, Margaret. 1975. *Introduction to systemic linguistics I*. London: Batsford Ltd.

Benedikt, Michael. 1992. *Cyberspace: First steps*. Cambridge, Mass.: MIT Press.

Bennett, Charles H. 1973. Logical reversibility of computation. *IBM Journal of Research and Development* 6: 525–32.

———. 1982. The thermodynamics of computation—a review. *International Journal of Theoretical Physics* 21, no. 12: 905–39.

———. 1984. Thermodynamically reversible computation. *Physical Review Letters* 53, no. 12: 1202.

Brooks, Rodney. 1991a. Intelligence without reason. MIT Artificial Intelligence Laboratory, Memo 1293.

———. 1991b. New approaches to robotics. *Science* 253: 1227–32.

Brouwer, L. E. J. 1952. Historical background, principles, and methods of intutionism. *South African Journal of Science* 49: 139–46.

Brown, John Seely. 1991. Research that reinvents the corporation. *Harvard Business Review* 69: 102.

Buchler, Justus. 1940. *The philosophy of Peirce: Selected writings*. London: Routledge.

Changeux, J.-P. 1986. *Neuronal man: The biology of mind*. Oxford: Oxford University Press.

Changeux, J.-P., and Alain Connes. 1995. *Conversations on mind, matter, and mathematics*. Edited and translated by M. B. Debevoise. Princeton, N.J.: Princeton University Press.

Coleman, Edwin. 1988. *The role of notation in mathematics*. Ph.D. thesis. Dept. of Linguistics, University of Adelaide, Australia.

———. 1990. Paragraphy. *Information Design Journal* 6, no. 2: 131–46.

———. 1992. Presenting mathematical information. In *Designing infor-*

mation for people, edited by R. Penman and D. Sless, 43–56. Canberra: ANU Press.

Crary, Jonathan. 1988. Modernizing vision. In *Vision and visuality*, edited by Hal Foster, 29–44. Seattle, Wash.: Bay Press.

Culler, Jonathan. 1981. *The pursuit of signs: Semiotics, literature, deconstruction*. Ithaca, N.Y.: Cornell University Press.

Davis, Philip. 1993. Visual theorems. *Educational Studies in Mathematics* 24 (1993): 333–44.

Deleuze, Gilles. 1990. *The logic of sense*. Translated by Mark Lester. New York: Columbia University Press.

Deleuze, Gilles, and Félix Guattari. 1987. *A thousand plateaus*. Translated by Brian Massumi. Minneapolis: University of Minnesota Press.

———. 1994. *What is philosophy?* Translated by Hugh Tomlinson and Graham Burchell. New York: Columbia University Press.

Donald, Merlin. 1991. *Origins of the modern mind: Three stages in the evolution of culture*. Cambridge, Mass.: Harvard University Press.

Eco, Umberto. 1976. *A theory of semiotics*. London: Macmillan.

Edelman, Gerald M. 1987. *Neural Darwinism: The theory of neuronal group selection*. New York: Basic Books.

———. 1992. *Bright air, brilliant fire: On the matter of the mind*. New York: Basic Books.

Edelman, Gerald M., and V. B. Mountcastle. 1978. *The mindful brain*. Cambridge, Mass.: MIT Press.

Eddington, Arthur. 1935. *New pathways in science*. Cambridge: Cambridge University Press.

Ernest, Paul. 1998. *Social constructivism as a philosophy of mathematics*. Albany: State University of New York Press.

Feynman, Richard P. 1967. *The character of physical law*. Cambridge, Mass.: MIT Press.

———. 1982. Simulating physics with computers. *International Journal of Theoretical Physics* 21, nos. 6–7: 467–88.

Fredkin, Edward, and Tomasso Toffoli. 1982. Conservative logic. *International Journal of Theoretical Physics* 21, nos. 3–4: 219–53.

Frege, Gottlob. 1967. The thought: A logical enquiry. In *Philosophical logic*, edited by P. F. Strawson, 17–38. Oxford: Oxford University Press.

Friedberg, R., and T. D. Lee. 1983. Discrete quantum mechanics. *Nuclear Physics* B225: 1–21.

Gardner, Martin. 1989. Beauty in numbers. *New York Review of Books*, 16 Mar.

Gelernter, David. 1992. *Mirror worlds*. New York: Oxford University Press.

Gould, Stephen Jay. 1989. *Wonderful life*. New York: W. W. Norton.

Greenspan, Donald. 1973. *Discrete models*. Reading, Mass.: Addison Wesley.

Guattari, Félix. 1995. *Chaosmosis*. Translated by Paul Bains and Julian Pefanis. Bloomington: Indiana University Press.

Halliday, M. A. K. 1978. *Language as social semiotic*. London: Edward Arnold.

Hamilton, William R. 1853. *Lectures on quaternions*. Dublin: Hodges and Smith.

Haraway, Donna J. 1991. Situated knowledges: The science question in feminism and the privilege of partial perspective. In *Simians, cyborgs, and women: The reinvention of nature*, by D. J. Haraway, 183–201. New York: Routledge.

Harris, Roy. 1986. *The origin of writing*. London: Duckworth.

———. 1987. *The language machine*. London: Duckworth.

Hayles, N. Katherine. 1993. The materiality of informatics. *Configurations* 1: 147–60.

Hilbert, D., and P. Bernays. 1934. *Grundlagen der Mathematik 1*. Berlin: Springer Verlag.

Husserl, Edmund. 1981. The origin of geometry. Translated by David Carr. In *Husserl: Shorter works*, edited by P. McCormick and F. Ellison, 251–70. South Bend, Ind.: Notre Dame University Press.

Johnson, Mark. 1987. *The body in the mind: The bodily basis of meaning, imagination, and reason*. Chicago: University of Chicago Press.

Landauer, Rolf. 1961. Irreversibility and heat generation in the computing process. *IBM Journal of Research and Development* 3: 183–91.

———. 1986. Computation and physics: Wheeler's meaning circuit? *Foundation of Physics* 16, no. 6: 551–64.

———. 1991. Information is physical. *Physics Today* (May): 23–29.

Layzer, David. 1990. *Cosmogenesis: The growth of order in the universe*. New York: Oxford University Press.

Lee, T. D. 1983. Can time be a discrete dynamical variable? *Physics Letters* 122B, nos. 3–4 (Mar.): 217–20.

Lemke, Jay. 1998. Multiplying meaning: Visual and verbal semiotics in scientific text. In *Reading science*, edited by J. R. Martin and R. Veel, 87–113. London: Routledge.

Lenoir, Timothy, and Christophe Lécuyer. Forthcoming. *Visions of theory: Fashioning molecular biology as an information science*.

Lorenzen, P. 1955. *Einfuhrung in die operative Logik und Mathematik.* Berlin: Springer Verlag.
Lyotard, Jean-François. 1984. *The postmodern condition.* Minneapolis: University of Minnesota Press.
McCulloch, W. S. 1965. *Embodiments of mind.* Cambridge, Mass.: MIT Press.
Merleau-Ponty, Maurice. 1962. *The phenomenology of perception.* Translated by Colin Smith. New York and London: Humanities Press.
Mortensen, Chris, and Lesley Roberts. 1997. Semiotics and the foundations of mathematics. *Semiotica* 115: 1–25.
O'Halloran, Kay. 1996. *The discourses of secondary school mathematics.* Ph.D. thesis. Dept. of Education, Murdoch University, Australia.
Peirce, Charles Sanders. 1958. *Studies in the philosophy of Charles Sanders Peirce.* Vol. 5 of *Collected writings*, edited by Philip Wiener. Cambridge, Mass.: Harvard University Press.
Penrose, Roger. 1989. *The emperor's new mind.* London: Penguin Books.
Rotman, Brian. 1988. Toward a semiotics of mathematics. *Semiotica* 72, nos. 1–2 (1988): 1–35.
———. 1993a. *Ad infinitum . . . the ghost in Turing's machine: Taking God out of mathematics and putting the body back in.* Stanford, Calif.: Stanford University Press.
———. 1993b. *Signifying nothing: The semiotics of zero.* Stanford, Calif.: Stanford University Press.
———. 1996. Review of J.-P. Changeux and Alain Connes: *Conversations on mind, matter, and mathematics. Times Literary Supplement*, Jan.
———. 2000. Becoming beside oneself. *Configurations.*
Serres, Michel. 1977. *La naissance de la physique dans le texte de Lucrèce.* Paris: Minuit.
Shin, Sun-Joo. 1994. *The logical status of diagrams.* New York: Cambridge University Press.
Stephenson, Neal. 1992. *Snow crash.* New York: Bantam Books.
Suchman, Lucy. 1987. *Plans and situated actions: The problem of human-machine communication.* New York: Cambridge University Press.
Toffoli, Tommaso. 1982. Physics and computation. *International Journal of Theoretical Physics* 21, nos. 3–4: 65–175.
Varela, Francisco J., Evan Thompson, and Eleanor Rosch. 1991. *The embodied mind: Cognitive science and human experience.* Cambridge, Mass.: MIT Press.

Vera, A. H., and H. A. Simon. 1993. Situated action: A symbolic interpretation. *Cognitive Science* 17: 7–48.

Weber, Samuel. 1987. *Institution and interpretation*. Minneapolis: University of Minnesota Press.

West, Bruce J. 1985. *An essay on the importance of being linear*. New York: Springer Verlag.

Weyl, H. 1949. *Philosophy of mathematics and natural science*. Princeton, N.J.: Princeton University Press.

Wheeler, J. H. 1986. "Physics as meaning circuit": Three problems. In *Frontiers of nonequilibrium statistical physics*, edited by G. T. Moore and M. O. Scully, 25–32. New York: Plenum.

———. 1988. World as system self-synthesized by quantum networking. *IBM Journal of Research and Development* 32, no. 1: 4–15.

———. 1990. Information, physics, quantum: The search for links. In *Complexity, entropy, and the physics of information*, edited by Wojciech H. Zurek, 3–28. Redwood City, Calif.: Addison Wesley.

Wolfram, Stephen. 1985. Undecidability and intractability in theoretical physics. *Physical Review Letters* 54, no. 8 (25 Feb.): 735–38.

INDEX OF PERSONS

Aristotle, 72, 73, 121
Arnold, V. I., 58

Bateson, Gregory, 58
Benedikt, Michael, 63
Bennett, Charles, 86, 87, 92, 93
Berkeley, George, 73
Bernoulli, Johann II, 75
Berry, Margaret, 9
Bishop, Errett A., 129, 130
Bohr, Niels, 51
Bourbaki, Charles-Denis-Sauter, 56, 58
Brooks, Rodney, 112, 113–15
Brouwer, Jan, 1, 2, 47, 73, 129, 130; and intuitionism, 6, 25–28, 38, 80
Brown, John Seely, 117–19

Cantor, Georg, 73
Carnap, Rudolf, 3
Changeux, Jean-Paul: *Neuronal Man*, 111; *Conversations on Mind, Matter, and Mathematics*, 127
Cohen, Paul, 49, 73–74
Coleman, Edwin, 43
Connes, Alain: *Conversations on Mind, Matter, and Mathematics*, 127
Crary, Jonathan, 159n1
Culler, Jonathan, 108

Davis, Philip, 57–58
Dedekind, Julius Wilhelm Richard, ix
Deleuze, Gilles, 125, 150; nomadism, 137, 142, 144, 146, 151, 152, 160n7; *A Thousand Plateaus*, 138, 139–41
Derrida, Jacques, 45, 68, 108, 109
Descartes, René, 28, 100
Donald, Merlin, 68; *The Origins of the Modern Mind*, 106
Doyle, Richard, 100

Eddington, Arthur Stanley, 89, 134
Edelman, Gerald, 158n1; *Neural Darwinism*, 111–12
Einstein, Albert, 51
Epicurus, 150
Ernest, Paul, 43
Euclid, 129
Euler, Leonhard, 75

Feynman, Richard, 86; discrete physics of, 81–82, 83–84
Fourier, Jean-Baptiste-Joseph, 75
Fredkin, Edward, 86
Frege, Gottlob, 1, 2, 3, 6, 30, 38, 73; and realism, 31, 32–34
Freudenthal, Hans, 120

Galileo Galilei, 51
Gardner, Martin, 76–77, 157–58n1
Gelernter, David: *Mirror Worlds*, 65
Gödel, Kurt, 23, 49, 73, 80
Gould, Stephen Jay, 120
Greenspan, Donald, 79
Guattari, Félix, 125; nomadism, 137, 142, 144, 146, 151, 152, 153, 160n7; *A Thousand Plateaus*, 138, 139–41
Gutenberg, Johannes, 97

Halliday, M., 42
Hamilton, William Rowan, 135

Haraway, Donna, 121; "Situated Knowledges," 119
Harris, Roy, 45, 46, 58, 63
Hayles, N. Katherine, 109
Hegel, Georg Wilhelm Friedrich, 36
Hilbert, D., 1, 2, 26, 38; formalism of, 5–6, 21–22, 24–25, 44
Husserl, Edmund, 47–48, 56, 57, 61

Kripke, Saul, 3
Kronecker, Leopold, 38, 72, 129, 147

Landauer, Rolf, 83, 84–85, 86, 87; on continuum mathematics, 88–89, 90–91; and self-consistent theory, 92, 94–95, 96
Laplace, Pierre-Simon de, 76
Lécuyer, Christophe, 67
Lee, T. D., 80–81, 82
Leibniz, Gottfried Wilhelm, 120
Lemke, Jay, 42
Lenoir, Tim, 67
Lewontin, Richard C., 110
Lorenzen, P., 39
Lucretius, 125, 141

McCulloch, Warren: *Embodiments of Mind*, 109
Mach, Ernst, 51
McLuhan, Marshall, 45
Marx, Karl, 36
Maxwell, James Clerk, 75
May, Robert, 76
Mead, G. H., 116
Merleau-Ponty, 57, 158n2
Mill, John Stuart, 30
Mortensen, Chris, 43

Nagel, Thomas, 119
Neumann, John von, 77

O'Halloran, Kay, 42
Ong, Walter, 45
Oyama, Susan, 110–11

Parmenides, 72
Pascal, Blaise, 153
Peirce, Charles Sanders, 1, 4, 42, 53, 59, 148; on self-reflection, 13, 51; and semiotics, 15–16, 22–23, 49–50, 62, 122; on proofs as arguments, 17–18
Penrose, Roger, 82, 89, 90, 91, 157–58n1
Plato, 31, 121, 127
Poincaré, Jules-Henri, 76
Pythagoras, 127

Quine, W. V. O., 3, 7, 80

Roberts, Lesley, 43
Rosch, Eleanor: *The Embodied Mind*, 111, 158n2
Russell, Henry Norris, 3

Sagan, Carl, 120
Saussure, Ferdinand, 63, 109
Schrödinger, Erwin, 78
Serres, Michel, 141, 169n9
Skolem, Thoralf Albert, 49
Stephenson, Neal: *Snow Crash*, 65
Stevin, Simon: "The Dime," 73
Suchman, Lucy, 118; *Plans and Situated Action*, 115–17
Sun-Joo Shin, 43

Thompson, Evan: *The Embodied Mind*, 111, 158n2
Toffoli, Tommaso, 83, 86, 87, 94
Turing, Alan Mathison, 49, 77, 136

Varela, Francisco: *The Embodied Mind*, 111, 158n2

Weber, Samuel, 50, 62
West, Bruce, 75
Weyl, Herman, 15
Wheeler, John, 79–80, 81, 82, 83, 90
Wittgenstein, Ludwig Josef Johan, 3
Wolfram, Stephen, 83

Zeno, 72
Zurek, Wojciech H., 86

WRITING SCIENCE

Brian Rotman, *Mathematics as Sign: Writing, Imagining, Counting*

Thierry Bardini, *The Personal Interface: Douglas Engelbart and the Genesis of Personal Computing*

Lily E. Kay, *Who Wrote the Book of Life? A History of the Genetic Code*

Jean Petitot, Francisco J. Varela, Bernard Pachoud, and Jean-Michel Roy, eds., *Naturalizing Phenomenology: Issues in Contemporary Phenomenology and Cognitive Science*

Francisco J. Varela, *Ethical Know-How: Action, Wisdom, and Cognition*

Bernhard Siegert, *Relays: Literature as an Epoch of the Postal System*

Friedrich A. Kittler, *Gramophone, Film, Typewriter*

Dirk Baecker, ed., *Problems of Form*

Felicia McCarren, *Dance Pathologies: Performance, Poetics, Medicine*

Timothy Lenoir, ed., *Inscribing Science: Scientific Texts and the Materiality of Communication*

Niklas Luhmann, *Observations on Modernity*

Dianne F. Sadoff, *Sciences of the Flesh: Representing Body and Subject in Psychoanalysis*

Flora Süssekind, *Cinematograph of Words: Literature, Technique, and Modernization in Brazil*

Timothy Lenoir, *Instituting Science: The Cultural Production of Scientific Disciplines*

Klaus Hentschel, *The Einstein Tower: An Intertexture of Dynamic Construction, Relativity Theory, and Astronomy*

Richard Doyle, *On Beyond Living: Rhetorical Transformations of the Life Sciences*

Hans-Jörg Rheinberger, *Toward a History of Epistemic Things: Synthesizing Proteins in the Test Tube*

Nicolas Rasmussen, *Picture Control: The Electron Microscope and the Transformation of Biology in America, 1940–1960*

Helmut Müller-Sievers, *Self-Generation: Biology, Philosophy, and Literature Around 1800*

Karen Newman, *Fetal Positions: Individualism, Science, Visuality*

Peter Galison and David J. Stump, eds., *The Disunity of Science: Boundaries, Contexts, and Power*

Niklas Luhmann, *Social Systems*

Hans Ulrich Gumbrecht and K. Ludwig Pfeiffer, eds., *Materialities of Communication*